Econometric Analysis of Carbon Markets

Julien Chevallier

Econometric Analysis of Carbon Markets

The European Union Emissions Trading Scheme and the Clean Development Mechanism

 Springer

Dr. Julien Chevallier
CGEMP/LEDa
Department of Economics
University Paris Dauphine
Place du Marechal de Lattre de Tassigny
75016 Paris
France
julien.chevallier@dauphine.fr

ISBN 978-94-007-2411-2   e-ISBN 978-94-007-2412-9
DOI 10.1007/978-94-007-2412-9
Springer Dordrecht Heidelberg London New York

Library of Congress Control Number: 2011938668

Cover design: VTeX UAB, Lithuania

Printed on acid-free paper

Springer is part of Springer Science+Business Media (www.springer.com)

110017161

To my family.

Preface

The idea of writing this book naturally came after a discussion with my editor, whom I warmly thank for making this project possible. This book should be extremely useful for researchers and working professionals (trading managers, energy and commodity traders, quantitative analysts, consultants, utilities) in the fields of econometrics and carbon finance. We define carbon markets as the environmental markets created to regulate the emissions of greenhouse gases (including CO_2), such as the European Union Emissions Trading Scheme (EU ETS) and the Kyoto Protocol (more precisely the Clean Development Mechanism, CDM).

This book is intended for readers with a basic understanding of time series econometrics (such as the linear regression model, vector autoregression, and cointegration). Useful textbooks to refresh concepts on this matter are Gujarati (*Basic Econometrics*, McGraw-Hill) and Hamilton (*Time Series Analysis*, Princeton University Press). It can be used for teaching econometrics applied to carbon markets at undergraduate or postgraduate levels (M.Sc., MBA), and as a reference for professionals.

The content of this book has been presented during lectures on the econometrics of energy markets and on commodity finance at the University Paris Dauphine. It draws on feedback and practical exercises developed with former M.Sc. Students, to whom this book is likewise dedicated.

Through the analysis of the EU ETS and the CDM, the book shows how to use a variety of econometric techniques to analyze an evolving and expanding carbon market sphere worldwide. The book offers a mix of knowledge on emissions trading with practical applications to carbon markets. It covers the stylized facts on carbon markets from an economics perspective, as well as key aspects on pricing strategies, risk and portfolio management. On the one hand, it contains useful information on how to interpret the historical development of the carbon price (until the present time). On the other hand, it is instructive to teach students (advanced undergraduates, M.Sc., MBA) and researchers how to use these techniques to perform similar exercises as carbon markets evolve and expand. Therefore, these techniques may be re-used as new national and regional schemes appear in the near future of environmental regulation (China, USA, etc.).

Chapters 1 and 2 provide an accessible introduction to the fundamentals of time series econometrics to support anyone new to the material in learning the principles of carbon markets. Starting in Chap. 1 with the analysis of descriptive statistics for carbon prices, the econometrics learning process is regular throughout the book. Chapter 1 has a broad coverage, as the book starts with the review of international climate policies (and thus of national and regional carbon markets initiatives as in the US, Japan, China, Australia, etc.). In addition, it contains a brief introduction to the mechanism of emissions trading, and the main features of the European carbon market. It includes little econometric techniques (only a review of descriptive statistics), but it contains nonetheless useful background information for the book.

Chapter 2 deals with carbon price drivers (mainly institutional decisions, energy prices and extreme weather events) in the context of the linear regression model. It covers perhaps the most interesting topic concerning CO_2 prices, i.e. what are the relevant price fundamentals on carbon markets? Starting with the use of dummy variables and the Bai-Perron structural break test, it relies heavily on the linear regression model to show the influence of other energy prices. Weather events are accounted for by using threshold variables (above or below a given temperature). The Appendix contains a useful review of the multivariate GARCH modelling framework applied to energy and CO_2 prices.

Chapters 3 to 6 offer a smooth transition from being taught the basics in time series econometrics to actively research carbon markets for term-papers or dissertations for advanced undergraduate and/or postgraduate students (M.Sc., MBA). Chapter 3 deals with an equally important topic: what is the relationship between the newly created emissions allowances and the pre-existing macroeconomic environment? This topic is tackled first by looking at the link between the equity/bond markets and the emissions market in a GARCH modeling framework. Then, the broad links with macroeconomic, financial and commodity spheres are captured in factor models. Next, the link with industrial production is studied by resorting to nonlinearity tests, threshold models and Markov regime-switching.

Chapter 4 focuses on the Clean Development Mechanism, which may be seen as a proxy of 'world' carbon prices in the absence of post-Kyoto agreements. Besides the description of the contracts, this chapter details the use of vector autoregression, Granger causality, cointegration techniques (with the EU carbon price) and the Zivot-Andrews structural break test. Arbitrage strategies between the European market and the CDM are studied in a GARCH modeling framework with microstructure variables in order to explain the existence of the 'EUA-CER spread'. The Appendix illustrates the use of the Markov regime-switching models with EUAs and CERs.

Chapter 5 deals with risk-hedging strategies and portfolio management. The main risk factors related to carbon assets are summarized. The measure of risk premia in CO_2 spot and futures prices is detailed based on commodity markets models and linear regressions. Besides, carbon price risk management strategies are described by the means of an econometric analysis of the factors influencing fuel-switching in the power sector. Finally, portfolio management techniques with carbon prices are explained in details with standard mean-variance optimization techniques. The Appendix recalls how to use option prices for risk management by computing implied volatilities.

Chapter 6 contains more advanced econometric techniques. It investigates whether or not the volatility of carbon prices shifts as CO_2 futures contracts reach maturity. Three approaches are used as a proxy for volatility in order to shed light on this issue: GARCH models, net cost-of-carry, and realized volatility. The so-called Samuelson hypothesis is tested in a regression framework, by accounting for seasonality and liquidity issues. The Appendix covers statistical techniques that are useful in order to detect instability in the volatility of carbon prices.

In order to replicate part of the empirical applications detailed in the various chapters, this book recommends the use of the R software. R is a free software designed for statistical and econometric analysis. Its installation is extremely easy under any operating software, and comes with a documentation. The packages that are required to duplicate the results must be downloaded from the Comprehensive R Archive Network (CRAN) and pre-installed by the user with the command `install.packages()` and by following the command prompt. Finally, the necessary data and R codes can be found on the author's website: http://sites.google.com/site/jpchevallier/publications/books/springer/.

This companion website with hyperlinks to data and computer codes related to the models, and the replication of examples discussed in the book (where it is relevant) ensure that the reader is able to develop his/her programming skills. Besides, this data warehouse with computer codes constitutes an original means of disseminating knowledge. The interested reader will also find a Frequently Asked Questions (FAQ) section on the website.

For developing R packages used in this book, I wish to thank Achim Zeileis, Friedrich Leisch, Bruce Hansen, Kurt Hornik, Christian Kleiber, Patrick Brandt, Bernhard Pfaff, Harald Schmidbauer, Angi Roesch, Vehbi Sinan Tunalioglu, Diethelm Wuertz, Yohan Chalabi, Michal Miklovic, Chris Boudt, Pierre Chausse, Douglas Bates, Ladislav Luksan, as well as the R core development team: http://www.r-project.org/.

In terms of course use, this book may be appealing to courses in finance, such as principles of financial markets, financial economics and financial econometrics. It will also be relevant to courses in energy, environmental and resource economics, as it covers the EU ETS and the Kyoto Protocol which constitute landmark environmental regulation policies. Since the carbon price constitutes a new commodity, it will also be eligible to courses on commodity markets and risk management on these markets. Finally, empirical applications may be re-used in exercises for standard (introductory) applied econometrics courses (such as time series analysis). Problem sets (with solutions manual) are presented at the end of the relevant chapters, and presentation slides for each chapter are available for instructors on the author's companion website.

Last but not least, I wish to thank many people with whom I have been interacting over the years, either in carbon finance or in academia, which led me to develop the skills and material necessary to write this book: Emilie Alberola, Benoit Cheze, Pascal Gastineau, Florian Ielpo, Maria Mansanet-Bataller, Morgan Herve-Mignucci, Benoit Leguet, Anais Delbosc, Raphael Trotignon, Benoit Sevi, Yannick Le Pen, Derek Bunn, ZongWei Luo, ZhongXiang Zhang, Denny Ellerman, Olivier Godard, Katheline Schubert, Natacha Raffin, Philippe Quirion, David Newbery, Fabien Roques, Sam Fankhauser, Simon Dietz, Eric Neumayer, Luca Taschini,

Ian Lange, Ibon Galarraga, Mikel Gonzalez, Ralf Antes, Bernd Hansjurgens, Peter Letmathe, Massimiliano Mazzanti, Anna Montini, Simon Buckle, Walter Distaso, Robert Kosowski, Jan Ahmerkamp, Andreas Loschel, Daniel Rittler, Waldemar Rotfuss, Emeric Lujan, Erik Delarue, William D'haeseleer, Valerie Mignon, Anna Creti, Gilles Rotillon, Pierre-Andre Jouvet, Johanna Etner and Christian De Perthuis. At the CGEMP of the University Paris Dauphine, I wish to thank all the team and more particularly Jean Marie Chevalier and Patrice Geoffron who helped me to focus on my research.

Paris, France   Julien Chevallier

Contents

1   Introduction to Emissions Trading 1
  1.1 Review of International Climate Policies 1
      1.1.1 From Rio to Durban 1
      1.1.2 The Burgeoning EU CO_2 Allowance Trading Market . . . 3
  1.2 Market Design Issues 4
      1.2.1 Initial Allocation Rules 4
      1.2.2 Equilibrium Permits Price 5
      1.2.3 Spatial and Temporal Limits 6
      1.2.4 Safety Valve 10
  1.3 Key Features of the EU Emissions Trading Scheme 10
      1.3.1 Scope and Allocation 10
      1.3.2 Calendar 12
      1.3.3 Penalties 12
      1.3.4 Market Players 12
  1.4 EUA Price Development 13
      1.4.1 Structure and Main Features of EU ETS Contracts 13
      1.4.2 Carbon Price 14
      1.4.3 Descriptive Statistics 15
      References 16

2 CO_2 Price Fundamentals 19
  2.1 Institutional Decisions 19
      2.1.1 Dummy Variables 19
      2.1.2 Structural Breaks 26
  2.2 Energy Prices 29
      2.2.1 Literature Review 30
      2.2.2 Oil, Natural Gas and Coal 31
      2.2.3 Electricity Variables 35
  2.3 Extreme Weather Events 40
      2.3.1 Relationship Between Temperatures and Carbon Prices . . 41
      2.3.2 Empirical Application 43

Appendix   BEKK MGARCH Modeling with CO_2 and Energy Prices .   45
Problems   48
References   52

3   Link with the Macroeconomy   55
3.1   Stock and Bonds Markets   55
      3.1.1   GARCH Modeling of the Carbon Price   55
      3.1.2   Relationship with Stock and Bond Markets   60
3.2   Macroeconomic, Financial and Commodity Indicators   66
      3.2.1   Extracting Factors Based on Principal Component Analysis   67
      3.2.2   Factor-Augmented VAR Analysis Applied to EUAs   68
3.3   Industrial Production   71
      3.3.1   Data   71
      3.3.2   Nonlinearity Tests   74
      3.3.3   Self-exciting Threshold Autoregressive Models   78
      3.3.4   Comparing Smooth Transition and Markov-Switching
              Autoregressive Models   87
References 101

4   The Clean Development Mechanism 105
4.1   CERs Contracts and Price Development 105
4.2   Relationship with EU Emissions Allowances 107
      4.2.1   VAR Analysis 107
      4.2.2   Cointegration 117
4.3   CERs Price Drivers 123
      4.3.1   Zivot-Andrews Structural Break Test 124
      4.3.2   Regression Analysis 126
4.4   Arbitrage Strategies: The CER-EUA Spread 131
      4.4.1   Why So Much Interest in This Spread? 131
      4.4.2   Spread Drivers 132
Appendix   Markov Regime-Switching Modeling with EUAs and CERs 133
Problems 138
References 145

5   Risk-Hedging Strategies and Portfolio Management 147
5.1   Risk Factors 147
      5.1.1   Idiosyncratic Risks 148
      5.1.2   Common Risk Factors 149
5.2   Risk Premia 151
      5.2.1   Theory on Spot-Futures Relationships in Commodity
              Markets 151
      5.2.2   Bessembinder and Lemmon's (2002) Futures-Spot
              Structural Model 153
      5.2.3   Empirical Application 154
5.3   Managing Carbon Price Risk in the Power Sector 157
      5.3.1   Economic Rationale 157
      5.3.2   UK Power Sector 158

        5.3.3   Factors Influencing Fuel-Switching 159
        5.3.4   Econometric Analysis 160
        5.3.5   Empirical Results 164
        5.3.6   Summary 169
  5.4   Portfolio Management 170
        5.4.1   Composition of the Portfolio 170
        5.4.2   Mean-Variance Optimization and the Portfolio Frontier . . 171
  Appendix   Computing Implied Volatility from Option Prices 173
       Problems 174
       References 177

**6   Advanced Topics: Time-to-Maturity and Modeling the Volatility of
Carbon Prices** 181
  6.1   The Relationship Between Volatility and Time-to-Maturity in
       Carbon Prices 181
  6.2   Background on the Samuelson Hypothesis 183
  6.3   Data 184
        6.3.1   Daily Frequency 184
        6.3.2   Intraday Frequency 185
  6.4   The 'Net Carry Cost' Approach 186
        6.4.1   Computational Steps 187
        6.4.2   Regression Analysis 188
        6.4.3   Empirical Results 189
  6.5   GARCH Modeling 190
        6.5.1   GARCH Specification 190
        6.5.2   Empirical Results 191
  6.6   Realized Volatility Modeling 192
        6.6.1   Computational Steps 192
        6.6.2   Regression Analysis 193
        6.6.3   Empirical Results 194
        6.6.4   Sensitivity Tests 195
  6.7   Summary 200
  Appendix   Statistical Techniques to Detect Instability in the Volatility
       of Carbon Prices 201
       References 206

Solutions 209
  A.1   Problems of Chap. 2 209
  A.2   Problems of Chap. 4 210
  A.3   Problems of Chap. 5 212

Index 213

Acronyms

ARCH	AutoRegressive Conditional Heteroskedasticity Model
ARIMA	AutoRegressive Integrated Moving Average Process
ARMA	AutoRegressive Moving Average Process
BEKK	Baba-Engle-Kraft-Kroner Multivariate GARCH Model
BERR	UK Department for Business, Enterprise and Regulatory Reform
BHHH	Berndt, Hall, Hall and Hausman optimization algorithm
BIS	UK Department for Business, Innovation and Skills
CAPM	Capital Asset Pricing Model
CDM	Clean Development Mechanism
CDM EB	Clean Development Mechanism Executive Board
CER	Certified Emissions Reductions
CH_4	Methane
CCGT	Combined Cycle Gas Turbines
CGARCH	Component Generalized AutoRegressive Conditional Heteroskedasticity Model
CITL	Community Independent Transactions Log
COP	Conferences of the Parties of the Kyoto Protocol
CPRS	Australian Carbon Pollution Reduction Scheme
CRB	Reuters/Jefferies Commodity Research Bureau Futures index
DECC	UK Department for Energy and Climate Change
DG CLIMA	Directorate-General for Climate Action
ECB	European Central Bank
EC	European Commission
ECM	Error-Correction Model
ECX	European Climate Exchange
EEX	European Energy Exchange
EGARCH	Exponential Generalized AutoRegressive Conditional Heteroskedasticity Model
EITE	Emission Intensive and Trade-Exposed companies
EUA	European Union Allowance
EU ETS	European Union Emissions Trading Scheme

FAVAR	Factor-Augmented VAR Model
FTSE	Financial Times Stock Exchange
GARCH	Generalized AutoRegressive Conditional Heteroskedasticity Model
GED	Generalized Error Distribution
GHG	Greenhouse Gas
GSCI	Goldman Sachs Commodity Indicator
HFC	Hydrofluorocarbon
HFC-23	Trifluoromethane
ICE	Intercontinental Exchange
IET	International Emissions Trading
IPCC	Intergovernmental Panel on Climate Change
IRF	Impulse Response Function
ITL	International Transactions Log
ITR	Intertemporal Exchange Ratio
IV	Implied Volatility
JI	Joint Implementation
LM	Lagrange Multiplier test statistic
LR	Likelihood Ratio test statistic
LULUCF	Land Use, Land Use Change and Forestry projects
MAC	Marginal Abatement Cost
MCMC	Markov Chain Monte Carlo Methods
MDH	Mixture of Distribution Hypothesis
MGARCH	Multivariate GARCH
MGGRA	US Midwestern Greenhouse Gas Reduction Accord
MOP	Meeting of the Parties of the Kyoto Protocol
MPCCC	Australian Parliament's Multi-Party Climate Change Committee
NAP	National Allocation Plan
NBP	National Balancing Point for UK gas spot price
NCC	Net carry cost
N_2O	Nitrous Oxide
NZ ETS	New Zealand emissions trading scheme
NZU	New Zealand Unit
OTC	Over-The-Counter trading
PARCH	Power AutoRegressive Conditional Heteroskedasticity Model
PCA	Principal Component Analysis
PFC	Perfluorocarbon
QML	Quasi Maximum Likelihood
RCM	Regime Classification Measure
RGGI	US Regional Greenhouse Gas Initiative
RV	Realized Volatility
SETAR	Self-Exciting Threshold Autoregressive Model
SSR	Sum of Squared Residuals
STAR	Smooth Transition Autoregressive Model
T-Bill	U.S. Treasury-Bill rate
TGARCH	Threshold Generalized AutoRegressive Conditional Heteroskedasticity Model

VAR	Vector Autoregression
VECM	Vector Error-Correction Model
UNFCCC	United Nations Framework Convention on Climate Change
WCI	US Western Regional Carbon Action Initiative

List of Boxes

Box 1.1   Normality Tests 15
Box 2.1   Stationarity and Unit Root Tests 20
Box 2.2   ARIMA(p, d, q) Modeling of Time Series 22
Box 2.3   The Ljung-Box-Pierce Residual Autocorrelation Test 26
Box 2.4   Granger Causality Test 33
Box 2.5   Homoskedasticity Tests 45
Box 3.1   ARCH(q) and GARCH(p, q) Models 56
Box 3.2   Further GARCH(p, q) Models 64
Box 4.1   Portmanteau Test of a VAR(p) 109
Box 4.2   Multivariate Jarque-Bera Statistic 111
Box 4.3   Multivariate ARCH Test 112
Box 4.4   OLS-Based CUSUM Processes 114
Box 5.1   Measuring Portfolio Performance with the Sharpe Ratio 171

List of Figures

Fig. 1.1   EUA: Raw data (*left panel*) and logreturns (*right panel*) from
           March 09, 2007 to March 31, 2009   14
Fig. 2.1   Diagnostic plots for the model estimated in Eq. (2.3)   25
Fig. 2.2   Autocorrelation and partial-autocorrelation functions for the
           residuals from the model estimated in Eq. (2.3)   25
Fig. 2.3   EUA: RSS and BIC for models with up to sixteen breaks   28
Fig. 2.4   EUA with three breaks (*vertical lines*) and 95% confidence
           intervals (*horizontal lines*)   29
Fig. 2.5   Brent: Raw data (*left panel*) and logreturns (*right panel*) from
           March 09, 2007 to March 31, 2009   32
Fig. 2.6   Natural Gas: Raw data (*left panel*) and logreturns (*right panel*)
           from March 09, 2007 to March 31, 2009   32
Fig. 2.7   Coal: Raw data (*left panel*) and logreturns (*right panel*) from
           March 09, 2007 to March 31, 2009   32
Fig. 2.8   Autocorrelation and partial-autocorrelation functions for the
           residuals from the model estimated in Eq. (2.9)   35
Fig. 2.9   Clean Dark Spread: Raw data (*left panel*) and logreturns (*right
           panel*) from March 09, 2007 to March 31, 2009   37
Fig. 2.10 Clean Spark Spread: Raw data (*left panel*) and logreturns (*right
           panel*) from March 09, 2007 to March 31, 2009   38
Fig. 2.11 Switch Price: Raw data (*left panel*) and logreturns (*right panel*)
           from March 09, 2007 to March 31, 2009   38
Fig. 2.12 Autocorrelation and partial-autocorrelation functions for the
           residuals from the model estimated in Eq. (2.16)   41
Fig. 2.13 Weather Events: European Temperatures Index (*left panel*) and
           Temperatures Deviation from Seasonal Average (*right panel*) in
           Degree Celsius from March 09, 2007 to March 31, 2009   42
Fig. 2.14 Autocorrelation and partial-autocorrelation functions for the
           residuals from the model estimated in Eq. (2.19)   44
Fig. 2.15 Residual series for the BEKK MGARCH model estimated in
           Eq. (2.23)   48

Fig. 2.16   Raw time-series (*left panel*) and logreturns (*right panel*) of the
            ECX EUA Futures Contract   49
Fig. 2.17   ACF functions of the ECX EUA Futures contract in raw
            (*left panel*) and logreturn (*right panel*) forms   49
Fig. 3.1   GARCH diagnostic plots (from *left* to *right* and *top* to *bottom*):
            EUA Time Series, Conditional Standard Deviation, ACF of
            Observations, ACF of Squared Observations, Cross Correlation,
            Residuals, Standardized Residuals, ACF of Standardized
            Residuals, ACF of Squared Standardized Residuals, QQ Plot of
            Standardized Residuals   59
Fig. 3.2   Macroeconomic risk factors for stock markets (from *left* to
            right and *top* to *bottom*): Euronext 100, dividend yields on the
            Euronext 100, Dow Jones Euro Stoxx 50, dividend yields on the
            Dow Jones Euro Stoxx 50   61
Fig. 3.3   Macroeconomic risk factors for bond markets (from *left* to
            right and *top* to *bottom*): Moody's corporate bonds rated AAA,
            Moody's corporate bonds rated BAA, junk bond yield for
            Moody's BAA-AAA bonds, Moody's speculative grade bonds,
            junk bond yield for Moody's speculative grade bonds   62
Fig. 3.4   Macroeconomic risk factors for interest rates (from *left* to *right*
            and *top* to *bottom*): 90-day US Treasury yield curve rate, return
            on the 90-day US Treasury yield curve rate, 2-year Euro Area
            Government Benchmark bonds yield, return on the 2-year Euro
            Area Government Benchmark bonds yield   63
Fig. 3.5   Macroeconomic risk factors for commodity markets (from
            left to *right* and *top* to *bottom*): Reuters-CRB Futures index,
            excess return on the Reuters-CRB Futures index, Goldman
            Sachs commodity indicator, excess return on the Goldman Sachs
            commodity indicator   64
Fig. 3.6   EUA Spot and Futures Prices from April 2008 to June 2009   68
Fig. 3.7   Factor loadings extracted from the large dataset of
            macroeconomic and financial (*left panel*) and commodity (*right
            panel*) variables with the PCA   69
Fig. 3.8   IRF analysis of carbon futures (*left panel*) and spot (*right panel*)
            prices following a recessionary shock in the FAVAR model   70
Fig. 3.9   EU 27 Industrial Production Index (*top*) and EUA Futures Price
            (*bottom*) in raw (*left panel*) and logreturn (*right panel*) forms
            from January 2005 to July 2010   72
Fig. 3.10   OLS-CUSUM Test for EU27PRODINDRET (*left panel*) and
            EUAFUTRET (*right panel*)   74
Fig. 3.11   Thresholds Estimated by the SETAR Models for the EU 27
            Industrial Production Index (*top panel*) and EUA Futures Price
            (*bottom panel*). Note: *Solid* (*open*) *circles* indicate data in the
            lower (*upper*) regime of a fitted threshold autoregressive model . .   86

Fig. 3.12   Model Diagnostics of the SETAR$(2, 4, 1)$ Model with $d = 3$ for
            EU27INDPRODRET   88
Fig. 3.13   Model Diagnostics of the SETAR$(2, 1, 4)$ Model with $d = 3$ for
            EUAFUTRET   89
Fig. 3.14   Transition function of the STAR model as a function of the
            observations for the transition variables *EU27INDPRODRET*
            (*left panel*), and *EU27INDPRODRET*(-1) (*right panel*)   96
Fig. 3.15   Smoothed transition probabilities estimated from the univariate
            Markov-switching model   99
Fig. 3.16   Regime transition probabilities estimated from the univariate
            Markov-switching model 100
Fig. 4.1   CER: Raw data (*left panel*) and logreturns (*right panel*) from
            March 09, 2007 to March 31, 2009 106
Fig. 4.2   VAR EUA-CER: Diagram of fit (*upper panels*) and residuals
            (*lower panels*) for the EUA Equation 111
Fig. 4.3   VAR EUA-CER: Diagram of fit (*upper panels*) and residuals
            (*lower panels*) for the CER Equation 112
Fig. 4.4   VAR EUA-CER: OLS-CUSUM Test for EUA (*left panel*) and
            CER (*right panel*) 114
Fig. 4.5   VAR EUA-CER: Impulse response functions for CER $\rightarrow$ EUA
            (*left panel*) and EUA $\rightarrow$ CER (*right panel*) 116
Fig. 4.6   VAR EUA-CER: Forecast Error Variance Decomposition for
            EUA (*upper panel*) and CER (*lower panel*) 117
Fig. 4.7   VECM EUA-CER: Residuals analysis for the EUA Equation . . . 124
Fig. 4.8   VECM EUA-CER: Residuals analysis for the CER Equation . . . 125
Fig. 4.9   Zivot-Andrews test statistic for the CER price series 126
Fig. 4.10   Diagnostic plots for the model estimated in Eq. (4.24) 129
Fig. 4.11   Autocorrelation and partial-autocorrelation functions for the
            residuals from the model estimated in Eq. (4.11) 129
Fig. 4.12   CER-EUA Spread: Raw data (*left panel*) and logreturns (*right
            panel*) from March 09, 2007 to March 31, 2009 131
Fig. 4.13   CER-EUA Spread drivers: Trading volumes (in 1,000 ton
            of CO_2) for EUAs (*left panel*) and CERs (*right panel*) 133
Fig. 4.14   CER-EUA Spread drivers: Market Momentum (*left panel*) and
            VIX Index (*right panel*) 133
Fig. 4.15   Smoothed transition probabilities estimated from the two-regime
            Markov-switching VAR for EUAs and CERs 136
Fig. 4.16   Regime transition probabilities estimated from the two-regime
            Markov-switching VAR for EUAs and CERs 136
Fig. 4.17   Raw time-series (*left panel*) and natural logarithms (*right panel*)
            of the ECX EUA December 2008/2009 Futures and Reuters CER
            Price Index from March 9, 2007 to March 31, 2009 138
Fig. 4.18   Logreturns of the ECX EUA December 2008/2009 Futures and
            Reuters CER Price Index from March 9, 2007 to March 31, 2009 . 140

Fig. 4.19   Impulse responses from EUA to CER (*left panel*) and from CER
            to EUA (*right panel*) Variables of the VAR(4) Model 143
Fig. 4.20   GARCH(1, 1) model autocorrelation of squared logreturns of the
            CER-EUA Spread 143
Fig. 4.21   GARCH(1, 1) model of the CER-EUA Spread innovations
            (*top panel*), conditional standard deviations (*middle panel*) and
            standardized innovations (*bottom panel*) 144
Fig. 5.1   Carbon Spot and Futures Prices (*top*) and convenience yield
            (*bottom*) from January 2 to December 15, 2008. Source:
            Bluenext, European Climate Exchange, Euribor 152
Fig. 5.2   Diagnostic plots for the model estimated in Eq. (5.6) 156
Fig. 5.3   Autocorrelation and partial-autocorrelation functions for the
            residuals from the model estimated in Eq. (5.6) 157
Fig. 5.4   Coal, natural gas and nuclear use in power generation in the UK
            (1970–2008). Source: UK Department for Energy and Climate
            Change (DECC) 159
Fig. 5.5   NBP gas daily spot price (2005–2008). Source: Reuters 161
Fig. 5.6   ARA coal daily spot price (2005–2008). Source: Reuters 161
Fig. 5.7   ECX carbon price (2005–2008). Source: ECX 162
Fig. 5.8   UK daily fuel-switching price (2005–2008). Source: calculation
            from the author by using Reuters, ECX 163
Fig. 5.9   Portfolio frontier analysis 172
Fig. 5.10   Standardized innovations (*left panel*) and conditional standard
            deviations (*right panel*) of the GARCH(p, q) model 175
Fig. 6.1   BlueNext CO_2 spot prices valid for Phase I and Phase II and
            ECX EUA December 2008 and December 2009 futures contracts
            from April 22, 2005 to December 15, 2009. Source: BlueNext,
            European Climate Exchange (ECX) 185
Fig. 6.2   Net carry cost (NCC) for the December 2008 (*top panel*) and
            December 2009 (*bottom panel*) carbon futures contracts from
            February 26, 2008 to December 15, 2009. Source: BlueNext,
            European Climate Exchange (ECX) 188
Fig. 6.3   Volatility signature plot for the ECX December 2008 futures
            contract from January 2, 2008 to December 15, 2008 with
            sampling frequencies ranging from 60 to 1,500 seconds 193
Fig. 6.4   Daily realized volatility of the ECX December 2008 (*top
            panel*) and December 2009 (*bottom panel*) futures contracts
            estimated with a 15-minute sampling frequency from January 2
            to December 15 for each year 194
Fig. 6.5   Volumes exchanged (in 1,000 ton of CO_2) in intraday data for the
            ECX December 2008 (*top panel*) and December 2009 (*bottom
            panel*) futures contracts from January 2 to December 15 for each
            year 198
Fig. 6.6   Conditional standard deviation for EUAs extracted from the
            EGARCH model estimated in Eq. (6.11) 201

Fig. 6.7   OLS- (*left panel*) and Recursive- (*right panel*) based CUSUM
            processes for the EGARCH model 203
Fig. 6.8   *F*-Statistic (*left panel*) and *p*-value (*right panel*) for structural
            breaks in the EGARCH model 204
Fig. 6.9   Monitoring EUA data with recursive (*left panel*) and moving
            (*right panel*) estimates tests for the EGARCH model 205

List of Tables

Table 1.1 Descriptive statistics for the carbon price in raw (EUA) and logreturns (EUARET) transformation 16
Table 2.1 Dummy Variables 21
Table 2.2 R Code: Linear regression with dummy variables 23
Table 2.3 R Output: Linear regression with dummy variables 24
Table 2.4 R Code: Bai-Perron Structural Break Test 27
Table 2.5 Cross-correlations between carbon (EUA) and energy (BRENT, NGAS, COAL) variables in logreturns transformation 32
Table 2.6 R Code: Linear Regression Model with Energy Variables 34
Table 2.7 R Output: Linear Regression Model with Energy Variables . . . 34
Table 2.8 Cross-correlations between carbon (EUA) and electricity (DARK, SPARK, SWITCH) variables in logreturns transformation 39
Table 2.9 R Code: Linear Regression Model with Electricity Variables . . 40
Table 2.10 R Output: Linear Regression Model with Electricity Variables . . 40
Table 2.11 R Code: Linear Regression Model with Extreme Weather Events 43
Table 2.12 R Output: Linear Regression Model with Extreme Weather Events 44
Table 2.13 R Code: Estimating the BEKK MGARCH model 46
Table 2.14 R Output: Estimating the BEKK MGARCH model 47
Table 2.15 Unit root tests for the EUA_t variable 50
Table 2.16 Unit root tests for the $D(EUA_t)$ variable 51
Table 2.17 Estimation of the mean-reversion model 52
Table 2.18 Diagnostic tests of the mean-reversion model 52
Table 3.1 R Code: GARCH Modeling 57
Table 3.2 R Output: GARCH coefficient estimates 58
Table 3.3 Descriptive statistics 73
Table 3.4 Unit root tests 73
Table 3.5 Keenan (1985) and Tsay (1986) nonlinearity tests results for $EU27INDPRODRET$ 77

Table 3.6   Keenan (1985) and Tsay (1986) nonlinearity tests results for
            EUAFUTRET   77
Table 3.7   Brock et al. (1987, 1996) test results for independence and
            identical distribution for *EU27INDPRODRET*   77
Table 3.8   Brock et al. (1987, 1996) test results for independence and
            identical distribution for *EUAFUTRET*   78
Table 3.9   Summary of Nonlinearity Tests Results   78
Table 3.10   LR test for threshold nonlinearity for *EU27INDPRODRET* with
            $p = 6$ 80
Table 3.11   LR test for threshold nonlinearity for *EUAFUTRET* with $p = 1$.   80
Table 3.12   Bootstrap p-values of threshold unit root tests (Caner and
            Hansen 2001)   82
Table 3.13   AIC of the SETAR Models Fitted to *EU27INDPRODRET*   83
Table 3.14   AIC of the SETAR Models Fitted to *EUAFUTRET*   83
Table 3.15   Fitted SETAR$(2, 4, 1)$ Model with $d = 3$ for
            EU27INDPRODRET   84
Table 3.16   Fitted SETAR$(2, 1, 4)$ Model with $d = 3$ for *EUAFUTRET* . . .   85
Table 3.17 p-values of the linearity test against STAR modelling   93
Table 3.18   Grid search results for the starting values of γ and c   93
Table 3.19   Estimation results of the *LSTAR*1 model with the transition
            variable *EU27INDPRODRET*   94
Table 3.20   Estimation results of the *LSTAR*1 model with the transition
            variable *EU27INDPRODRET*(-1)   95
Table 3.21 p-values of the test of no additive nonlinearity in STAR models .   96
Table 3.22 p-values of parameter constancy tests in STAR models   96
Table 3.23   Test of no error autocorrelation in STAR models up to lag 8 . . .   97
Table 3.24   Estimation results of the univariate Markov-switching model . .   98
Table 4.1   Descriptive statistics for Certified Emissions Reductions in raw
            (CER) and logreturns (CERRET) transformation 106
Table 4.2   R Code: VAR EUA-CER 108
Table 4.3   R Output: Selection of the order of the VAR(p) 109
Table 4.4   R Output: VAR EUA-CER Estimates 110
Table 4.5   R Output: Diagnostic tests for the residuals of the
            VAR EUA-CER 113
Table 4.6   R Output: Granger Causality Tests for the VAR EUA-CER . . . 115
Table 4.7   R Code: Johansen Cointegration Tests 118
Table 4.8   R Code: Johansen Cointegration Test with Structural Shift 120
Table 4.9   R Output: Johansen Cointegration Test with Structural Shift . . . 121
Table 4.10   R Code: VECM 122
Table 4.11   R Output: VECM 122
Table 4.12   R Output: VECM EUA-CER Diagnostic Tests 123
Table 4.13   R Code: Zivot-Andrews Structural Break Test for CERs 126
Table 4.14   R Output: Zivot-Andrews Structural Break Test for CERs 127
Table 4.15   Cross-correlations between CER price drivers in logreturns
            transformation 128

Table 4.16   R Code: Linear Regression Model for CERs 130

Table 4.17   R Output: Linear Regression Model for CERs 130

Table 4.18   Descriptive statistics for the CER-EUA Spread in raw
(SPREAD) and returns (SPREADRET) transformation 132

Table 4.19   Estimation results of the two-regime Markov-switching VAR
for EUAs and CERs 135

Table 4.20   Diagnostic tests of the two-regime Markov-switching VAR for
EUAs and CERs 137

Table 4.21   Summary Statistics for the Raw Time Series and Natural
Logarithms 139

Table 4.22   Cointegration Rank: Maximum Eigenvalue Statistic 139

Table 4.23   Cointegration Rank: Trace Statistic 139

Table 4.24   Cointegration Vector 139

Table 4.25   Model Weights 140

Table 4.26   VECM with $r = 1$ 140

Table 4.27   Summary Statistics for the Log-Returns 141

Table 4.28   VAR(4) Model Results for the EUA Variable 141

Table 4.29   VAR(4) Model Results for the CER Variable 142

Table 4.30   Diagnostic tests of VAR(4) Model 142

Table 4.31   Granger Causality Tests 143

Table 4.32   GARCH estimates for the CER-EUA Spread 144

Table 5.1   R Code: Estimating Bessembinder and Lemmon's (2002)
Structural Model 155

Table 5.2   R Output: Estimating Bessembinder and Lemmon's (2002)
Structural Model 155

Table 5.3   Descriptive statistics 163

Table 5.4   Unit root tests for the $switch_t$ variable 165

Table 5.5   Estimation results of Eqs. (5.7) and (5.8) during the full-sample . 166

Table 5.6   Estimation results of Eqs. (5.7) and (5.8) without NBP during
the full-sample 167

Table 5.7   Estimation results of Eqs. (5.7) and (5.8) during the
sub-sample #1 168

Table 5.8   Estimation results of Eqs. (5.7) and (5.8) during the
sub-sample #2 168

Table 5.9   Descriptive statistics for CO_2 prices 173

Table 5.10   GARCH estimates for CO_2 prices 174

Table 5.11   EGARCH estimates for CO_2 prices 175

Table 5.12   TGARCH estimates for CO_2 prices 176

Table 6.1   Descriptive statistics 186

Table 6.2   Testing the 'negative covariance' hypothesis in the CO_2
allowance market using the net carry cost approach 189

Table 6.3   Testing the Samuelson hypothesis in the CO_2 allowance market
using the AR(1)-GARCH(1, 1) model 191

Table 6.4   Testing the Samuelson hypothesis in the CO_2 allowance market
using realized volatility 195

Table 6.5   Testing the Samuelson hypothesis in the CO_2 allowance
             market using realized volatility and spot price volatility as an
             exogenous regressor 196
Table 6.6   Testing the Samuelson hypothesis in the CO_2 allowance market
             using realized volatility with seasonal dummy variables and
             number of trades as exogenous regressors 199
Table 6.7   R Code: OLS/Recursive based CUSUM Processes 202
Table 6.8   R Code: F-Statistic for structural breaks 204
Table 6.9   R Code: Monitoring structural breaks 205

Chapter 1
Introduction to Emissions Trading

Abstract This chapter constitutes an introduction to emissions trading. First, we detail the latest developments in the field of international climate negotiations. Second, we introduce the main characteristics of emissions trading, be it in terms of spatial and temporal limits, initial allocation rules or price caps and price floors. Third, we provide a descriptive analysis of the price development of EU CO_2 allowances.

1.1 Review of International Climate Policies

This section proposes a review of current negotiations concerning climate policies, as well as an overview of the EU carbon market. More details about the Kyoto Protocol's Clean Development Mechanism can be found in Chap. 4.

1.1.1 From Rio to Durban

In 1992, the United Nations Framework Convention on Climate Change (UNFCCC) has officialized the need address climate change issues. In 1997, the Kyoto Protocol defines the responsibility of industrialized countries to reduce their greenhouse gases (GHG) emissions by 5% compared to 1990 levels. As such, the Kyoto protocol paved the way for the creation of emission trading schemes and new commodity markets in order to achieve these targets. The so-called 'carbon markets' allow participants (countries or companies) to buy/sell allowances in order to comply with the emissions cap. In 2001, the Marrakech Agreements provide to the Kyoto Protocol the economic tools to finance and achieve the emissions reduction objectives. In 2007, the Bali Roadmap prepares the Conferences of the Parties (COP) 15/Meetings of the Parties (MOP) 5 of the Kyoto Protocol, to be held in Copenhagen (Denmark) on December 2009. This roadmap includes four blocks of negotiations (mitigation, adaptation, technologies and finance) in order to establish a post-2012 international climate agreement. Without achieving this target, the Copenhagen Agreement was nevertheless signed jointly by developed (including the US) and developing countries. The Cancun 2010 Conference and the Durban 2011 Conference constitute the last pieces in the puzzle of international climate negotiations (see [5]).

J. Chevallier, *Econometric Analysis of Carbon Markets*,
DOI 10.1007/978-94-007-2412-9_1, © Springer Science+Business Media B.V. 2012

1.1.1.1 Emissions Trading in the US

At the federal level, the negotiations concerning the creation of a national emissions trading scheme are more or less stalled. In 2007, the Lieberman-Warner Climate Security Act constitutes the first proposition of legislation, blocked by the Senate in June 2008, concerning the creation of an emissions trading market. In July 2009, the Waxman-Markey American Clean Energy and Security Act failed to pass in the hands of the Congress. The Kerry-Boxer proposition was similarly abandoned in November 2009. As of today, the Kerry-Lieberman American Power Act, which aims at reducing emissions by 83% in 2050 compared to 2005 levels, has not been pushed forward by the Obama administration. This latest piece of legislation would concern 7,500 US power and industrial installations emitting more than 25,000 ton of CO_2 per year, with a cap set to 2 billion ton of CO_2 equivalent per year. Such a scheme would replace existing regional initiatives.

Without a federal US emissions trading scheme, several regional initiatives are worth mentioning. In the Regional Greenhouse Gas Initiative (RGGI), ten North-Eastern states have been auctioning CO_2 allowances since January 2009. The price fluctuates in the range of $5 per ton of CO_2. Some additional states in the Mid-West and the West are planning to launch the Western Regional Carbon Action Initiative (WCI) in 2012. In this scheme, the emissions reduction target has been established to 15% by 2020 compared to 2005 levels. It follows the creation of the Midwestern Greenhouse Gas Reduction Accord (MGGRA) in 2007 with the purpose of creating an emissions trading system. Since 2010, the RGGI, WCI and MGGRA are cooperating in order to share their experience regarding the development of emissions trading, to inform the federal government concerning the development of climate policies, and to potentially enhance their collaboration in the near future. In California, the Bill AB32 entitled 'California Global Warming Solutions Act' was passed in 2006. Starting in 2012, the California Air Resources Board will create another regional emissions trading system. This scheme aims at reducing emissions by 80% by 2050 compared to 1990 levels. California is expected to be a larger market than the RGGI, with roughly 400 million ton covered by 2015. The 90% free allocation of this emissions trading scheme will stand in sharp contrast to the RGGI, where almost all allowances have been auctioned. Californian carbon prices are expected to fluctuate in the range of $10 to $15 per ton of CO_2 in 2012.

1.1.1.2 Emissions Trading in Australia and New-Zealand

In New Zealand, the government has voted the creation of a national emissions trading system in November 2009. Note that in 2008 the New Zealand emissions trading scheme (NZ ETS) concerned the forestry sector only. Since July 2010, this scheme concerns the energy, transport and industrial sectors. Waste and agriculture will be gradually included from 2011 to 2015. New Zealand Units (NZUs) are traded within this system. The import of international credits from other schemes is authorized. Therefore, emitters can buy NZUs from domestic suppliers, or Kyoto units. Note that emission intensive and trade-exposed (EITE) companies may apply for free allocation of NZUs. This comment also applies for some forest owners and fisheries. During the first three years of the scheme, a price cap of 25 NZ$ has been set along with a double accounting method (i.e. two quotas are surrendered in exchange of one ton of CO_2-equivalent emitted).

In Australia, the Carbon Pollution Reduction Scheme (CPRS, see [6]) has been rejected several times by the Senate. According to the current legislation proposition, the market mechanism put forth by the Australian parliament's Multi-Party Climate Change Committee (MPCCC) involves a fixed price start to an emissions trading scheme scheduled to commence in July 2012, subject to the successful passage of legislation in 2011.

1.1.1.3 Emissions Trading in Asia

In Japan, the government has set the goal to reduce emissions by 25% by 2020 compared to 1990 levels, thanks to renewable energy and emissions trading. The Japanese 'Basic Law on Global Warming' is unlikely to pass in 2011 given the recent earthquakes. As a consequence, we may expect Japan to introduce a national ETS in 2013 or 2014.

In South Korea, the government has adopted in December 2009 a 'Green Growth' Law which includes the creation of an emissions trading system. The goal is to start the scheme in 2012, but the negotiations are currently stalled with the industry. China is also discussing the introduction of a national carbon market by 2015. It will launch pilot emissions trading schemes in six cities and provinces (Beijing, Chongqing, Shanghai, Tianjin, Hubei, Guangdong) before 2013. China has pledged to reduce its carbon intensity by 40–45% by the end of 2020, compared to 2005 levels.

1.1.2 The Burgeoning EU CO₂ Allowance Trading Market

The European Union has set up an emissions trading system which concerns around 11,300 energy-intensive installations across the 27 Member-States. In the Kyoto Protocol, the EU has set its emissions reduction target to 8% by 2012 compared to 1990 levels. This goal has been converted into further emissions reduction objectives at the Member-State level following the so-called 'Burden Sharing Agreements'. On October 13, 2003, the Directive 2003/87/EC created the European Union Emissions Trading Scheme (EU ETS), which is a landmark environmental policy starting in January 2005. The scheme covers approximately half of GHG at the EU level.

The market has been set up according to three Phases. Phase I goes from 2005 to 2007, and corresponds to a 'warm-up' period prior to the introduction of the Kyoto Protocol. Phase II goes from 2008 to 2012, and is concomitant to the Kyoto Protocol. Phase III goes from 2013 to 2020. In absence of any post-2012 international climate policy framework to date, the third Phase corresponds to the objectives of the EU 'Energy-Climate Package' introduced in January 2008 to reduce emissions by 20% by 2020 (along with increasing energy efficiency by 20% and increasing the share of renewables in the energy mix to 20%, hence the '20-20-20' target).

The three main characteristics of the EU ETS may be summarized as follows. First, the allocation is (mostly) free during Phases I and II according to some pre-existing benchmark of emissions (i.e. 'grandfathering'). Phase III constitutes a radical shift to costly allowances, with the introduction of auctioning. Second, the EU ETS has been centralized at the national level for setting up National Allocation Plans (NAPs), and at the EU Commission level to harmonize its functioning. Namely, the EU Commission is in charge to administer the Community Independent Transaction Log (CITL), which oversees national registries and has been connected to the Kyoto Protocol's International Transaction Log (ITL). Third, the perimeter of the scheme is gradually expanding, with the inclusion of additional sectors such as petro-chemicals and aviation in 2012–2013. More details can be found in Sect. 1.3.

1.2 Market Design Issues

To regulate CO_2 emissions, the government may typically choose between a tax or a tradable permits program. In 1968, Dales [9] published the results of his work leading to tradable permits markets. The properties of tradable permits markets—which equalize marginal abatement costs among polluters—were then further studied theoretically by Montgomery (1972, [26]). This new environmental regulation tool was introduced progressively, for instance to control sulfur dioxide emissions in the US Title IV Clean Air Act of 1990.

In this section, we present practical issues regarding the creation of tradable permits markets related to (i) initial allocation rules, (ii) the equilibrium permits price, (iii) spatial and temporal limits, and (iv) price caps and price floors.

1.2.1 Initial Allocation Rules

When creating the tradable permits market, the task of the regulator consists, among other tasks, in setting a cap on emissions for N heterogenous polluting firms. The cap fixed by the regulator is noted $\overline{E}$, and may be further decomposed into individual permits allocation to firms $\overline{e}_i(t)$:

$$\sum_{i=1}^{N} \overline{e}_i(t) = \overline{E}(t) \qquad (1.1)$$

It is further assumed that firms will comply with the environmental constraint, either by spatial or intertemporal emissions trading.

Different allocation methodologies may be chosen, such as methodologies based on grandfathering (i.e. free distribution in proportion of recent emissions or a benchmark), auctioning, baseline emissions, or per capita allocation. It is not obvious to tell how tradable permits should be allocated. Even if theory suggests that auctioning is the best methodology [8, 19, 20], it may rise equity-based objections.

Newell et al. (2005, [27]) emphasize that tradable permits create rents, and grandfathering distributes those rents to existing firms while also erecting barriers to entry. To counterbalance these negative effects, let us note that the direct allocation of grandfathered permits offers a degree of political control over the distributional effects of regulation, thereby enabling the formation of majority coalitions. Once the permits market has been introduced, then the regulator may strengthen the environmental constraint.

The main interest of auctioning permits consists in the income transfer, which may take the form of a lump-sum or a tax rebate, and reduces pre-existing distortions. This recycling revenue effect, also called double dividend, takes place when auctioning permits allows decreasing taxes and reducing distortions. This should have a positive effect on supply of capital and labour, but the net effect on high- or low-income quintiles depends on the tax structure. If the tax cut benefits more to richer people, then there is a clear trade-off between efficiency and equity [10].

1.2.2 Equilibrium Permits Price

Let $C_i(e_i(t))$ be the abatement cost function[1] incurred by agent i in order to comply with his permits allocation $\bar{e}_i$. $C_i(e_i(t))$ is defined[2] on $\mathcal{R} \to \mathcal{R}$ and is of class $C^2[0; T]$. $C_i(e_i(t))$ is decreasing and convex in $e_i(t)$ with $C_i'(e_i(t)) < 0$, $C_i''(e_i(t)) > 0$ and $C_i(e_i(0)) = 0$.

Firms' i marginal abatement costs (MAC) are associated with a one-unit reduction from his emission level e_i at time t and are noted $-C_i'(e_i(t)) > 0$. At the equilibrium of a permits market in a static framework,[3] price-taking agents adjust emissions until the aggregated MAC is equal to the price P at time t:

$$P_t = -C_i'(e_i(t)) \tag{1.2}$$

---

[1] Compared to a situation where profits are unconstrained, abatement costs appear in order to meet the emission cap fixed by the regulator. See Leiby and Rubin (2001, [23]).

[2] Stated first by Montgomery (1972, [26]). The conditions given by Leiby and Rubin (2001, [23]) include the output $q(t)$ where $C_i[q_i(t), e_i(t), t]$ is strongly convex with $C_i'[q_i(t)] > 0$ and $C_i''[q_i(t)] > 0$. Properties of non-convex abatement cost functions may be found in Godby (2000, [14]).

[3] See Hahn (1984, [16]).

1.2.3 Spatial and Temporal Limits

Concerning temporal limits, banking and borrowing may be used to equalize marginal costs in present value. On tradable permits markets, *banking* refers to the possibility for agents to save unused permits for future use, while *borrowing* represents the possibility to borrow permits from future allocations for use in current period. By allowing agents to arbitrate between actual and expected abatement costs over specific periods, banking and borrowing permits form another dimension of flexibility, whereby agents can trade permits not only spatially but also through time.

Let $B_i(t)$ be the permits bank, with $B_i(t) > 0$ in case of banking and $B_i(t) < 0$ in case of borrowing. Any change in the permits bank is equal to the difference between $\overline{e}_i(t)$ and $e_i(t)$, respectively firm's i permits allocation and his emission level at time t. Setting the initial condition $B_i(0) = 0$, the banking and borrowing constraint may be written as:[4]

$$\dot{B}_i(t) = \overline{e}_i(t) - e_i(t) \tag{1.3}$$

1.2.3.1 Hotelling Conditions

Notwithstanding differences between a permit and an exhaustible resource, it is assumed in the literature that the Hotelling conditions for exhaustible resources must apply on a permits market. According to Liski and Montero (2005, [24]), the following differences may be highlighted between a tradable permits and a non-renewable resource. First, on a permits market with banking, the market may remain after the exhaustion of the bank; while the market of a non-renewable resource vanishes after the last unit extraction. Second, permits extraction and storage costs are equal to zero; while those costs are generally positive for a non-renewable resource. Third, the demand for an extra permit usually comes from a derived demand of other firms that also hold permits; while the demand for an extra unit of a non-renewable resource comes more often from a derived demand of another actor (e.g., a consumer). Consequently, the terminal and exhaustion conditions are introduced.

1.2.3.2 Terminal Condition

Let $[0, T]$ be the continuous time planning horizon. This planning period seems appropriate for a theoretical study of intertemporal emissions trading. For instance, Phase I of the EU ETS may be seen as a single period since the permits bank needs to be cleared by the end of 2007. Alternative time settings including distinct phases may be found in Schennach (2000, [30]) or Ellerman and Montero (2007, [13]), but they reflect the specific requirements of the US Acid Rain Program.

---

[4]See Rubin (1996, [29]).

At time T, cumulated emissions must be equal to the sum of each agent's depolluting objective, and therefore to the global cap $\overline{E}$ set by the regulator:[5]

$$\int_0^T \sum_{i=1}^N e_i(t)dt = \sum_{i=1}^N \overline{e}_i = \overline{E} \qquad (1.4)$$

1.2.3.3 Exhaustion Condition

At time T, there is no more permit in the bank (either stocked or borrowed):

$$\sum_{i=1}^N B_i(T) = 0 \qquad (1.5)$$

Those conditions ensure that agents gradually meet their depollution objective so that the marginal cost of depollution is equalized in present value over the time period, and the permits bank clears in the end. Note there is typically a truncation problem at the end of the period:

- if $B_i(T) > 0$, surplus allowances are worthless and agents are wasting permits;
- if $B_i(T) < 0$, agents need to pay a penalty.[6]

The authorization of these provisions appears desirable on a tradable permit market, since they allow firms to achieve their emissions reduction objective at least cost by smoothing their emissions overtime. However, they may also change the temporal profile as well as the magnitude of environmental damages. From the regulator's viewpoint, the best configuration of the intertemporal flexibility mechanism therefore consists in authorizing banking without restrictions, and in penalizing borrowing by using a non-unitary intertemporal exchange ratio (ITR, see the theoretical contributions by Rubin (1996, [29]), Kling and Rubin (1997, [21]), as well as the review by Chevallier (2011, [7])).

1.2.3.4 Economic Modelling

Banking and borrowing provisions can have a dramatic impact on the price path of CO_2 allowances. For instance, during Phase I of the EU ETS, the interdiction of banking and borrowing between 2007 and 2008 (i.e. the start of the Kyoto Protocol) led spot and futures prices for Phase I to zero (see [1] for more details on this topic). From 2008 onwards, there is unlimited banking and borrowing.

---

[5]See also Leiby and Rubin (2001, [23]).

[6]For instance, in the EU ETS the penalty is equal to 40 Euro plus one allowance to surrender during Phase I, and 100 Euro plus one allowance to surrender during Phase II.

Alberola and Chevallier (2009, [1]) have shown econometrically how banking restrictions have affected the price path of CO_2 allowances during 2005–2007 based on the following economic modelling. To characterize the equilibrium of the intertemporal allowances market during 2005–2007, their model is based on two strands of literature developed by Schennach (2000, [30]) and Slade and Thille (1997, [31]), which were first applied to the US SO_2 market by Helfand et al. (2006, [17]).

First, Schennach (2000, [30]) studies the banking behavior of regulated industrials in the US Acid Rain Program, and implicitly the behavior of spot prices in a stochastic, continous-time, infinite horizon model for allowance allocation, use and storage. Under certainty, the model predicts that the CO_2 price path would increase smoothly at the rate of interest according to the Hotelling rule.[7] Under uncertainty, the optimization program of risk-neutral agents is modelled as follows:

$$\begin{cases} \min_{e_t}\{E_0[\int_0^\infty e^{-\mu t} c_t(\varepsilon_t - e_t)dt]\} \\ \dot{S}_t = Y_t - e_t \\ S_t \geq 0 \end{cases} \tag{1.6}$$

with $E(t)$ a Von Neumann-Morgenstern expected utility function, e_t the emissions level after abatement, ε_t the counterfactual emissions level, $a_t = \varepsilon_t - e_t$ the total amount of abatement by all firms at time t, $c_t(a_t)$ the minimum total cost incurred by all firms to abate a_t, Y_t the total amount of allowances distributed to agents, S_t the number of allowances in the bank at time t, r the risk free interest rate, ρ the risk premium specific to holding allowances as an asset in a diversified portfolio of investments, and $\mu = r + \rho$ the rate specific the risky assets in the spirit of the capital asset pricing model (CAPM, see Markowitz (1959, [25]) and Sharpe (1966, [32])).

The solution to this problem is a continuous time version of the model by Pindyck (1993, [28]) of rational commodity pricing:

$$E_t[p_{t+1}] = (1 + \mu)p_t - \psi_t \tag{1.7}$$

with ψ_t a convenience yield.[8] Equation (1.7) therefore represents the basic economic modelling framework.

---

[7]According to Hotelling, the (real) price of an exhaustible resource over time rises at a percentage equal to the discount rate. This pioneering result in environmental economics shows that in an efficient exploitation of an exhaustible resource, the percentage change in net-price per unit of time should equal the discount rate in order to maximize the present value of the resource capital over the extraction period. See Chevallier (2011, [7]) for a review in the context of permits trading programs.

[8]According to Ellerman et al. (2000, [12]), an agent may benefit from holding a stock of allowances on hand to buffer itself against unexpected changes in emissions, which is called a convenience yield. More details can be found in Chap. 5.

Second, assuming an allowance may be considered as an exhaustible resource, the model of Slade and Thille (1997, [31]) provides an analogous theoretical framework by maximizing the function $V(R, p, \phi)$:

$$
\begin{cases}
\max_{q\tau} E_t\{\int_t^\infty e^{-\rho(\tau-t)}\pi_\tau d\tau\} \\
\dot{R}_\tau = -q\tau \\
R_\tau \geq 0, q\tau \geq 0 \\
\frac{\partial \phi}{\phi} = \mu_\phi dt + \sigma_\phi dz_\phi \\
\frac{\partial p}{p} = \mu_{pt} dt + \rho_p dz_p
\end{cases}
\tag{1.8}
$$

with $\pi_\tau = [p_\tau q_\tau - C(q_\tau, R_\tau, \phi_\tau)]$ the risk-adjusted profit at the discount rate ρ, ϕ a random productivity shock, $\dot{R}$ the state of the bank R as a function of the extraction rate q. The last two constraints represent a set of Ito processes with drift to model uncertainty.

At the equilibrium, the evolution of the allowance price P_t is:

$$
\frac{\frac{1}{\partial t} E_t \partial P_t}{P_t} = r + \beta(r^m - r) \equiv \rho
\tag{1.9}
$$

with E_t expected utility, r the risk-free interest rate, r^m the investment rate of return in a diversified portfolio, and β the risk premium specific to the asset. ρ represents the risk-adjusted discount rate used by firms to choose the emissions path that minimizes abatement costs. Therefore, we notice that at the equilibrium, the evolution of the allowance price p_t follows a Hotelling-CAPM relationship between the risk-free interest rate, the investment rate of return in a diversified portfolio, and the risk premium specific to the asset similar to (1.7).

Taken together, these economic models are useful to understand how the CO_2 price fluctuates in presence of banking and borrowing effects. Their application in an econometric analysis dedicated to the divorce between Phase I and Phase II carbon prices under the EU ETS may be found in Alberola and Chevallier (2009, [1]). Their results illustrate that the properties of the banking instrument highlighted above have been 'sacrificed' between 2007 and 2008.

1.2.3.5 Spatial Limits

Last but not least, concerning spatial limits, it is worth emphasizing scaling issues (Tietenberg (2006, [33])). Indeed, increasing the scale of the cap and trade system increases economic efficiency, but also decreases trade security. In addition, when setting up a tradable permits market, the regulator needs to take into account deposition constraints, by avoiding excedance of critical loads in specific geographical zones. Another concern lies in the proper design of national emission ceilings.

1.2.4 Safety Valve

A safety valve is an hybrid instrument designed to limit the cost of capping emissions at some target level, whereby the regulator offers to sell permits in whatever quantity at a predetermined price [33]. If prices are greater than expected, the marginal cost of abatement would be limited to the safety valve price. The regulator tries to set the emissions cap at a level where the expected marginal cost of meeting the constraint will match the beliefs about marginal benefits. Another advantage consists in keeping the attractiveness of a quantity target, and the use of a price mechanism in order to regulate the emissions of pollutants.

The main criticism of a safety valve consists in the determination of a 'fair' price by the regulator: if the safety valve price is too high, it will have no effect. Conversely, if the safety valve is too low, the quantity constraint is not binding anymore, and may be associated with a permanent tax. Moreover, there is a potential loss of 'environmental integrity', i.e. a fear of relaxing towards target reduction instead of supporting economically efficient implementation. The regulator shall thus attempt to avoid excessive violations of the original target. Finally, it is worth asking whether it appears useful to dilute the quantity constraint. For more details on price caps and price floors, see Grüll and Taschini (2011, [15]).

1.3 Key Features of the EU Emissions Trading Scheme

This section provides some additional details on the functioning of the EU ETS.

1.3.1 Scope and Allocation

The EU ETS has been created by the Directive 2003/87/CE. Across its 27 Member States, it covers large plants from CO_2-intensive emitting industrial sectors with a rated thermal input exceeding 20 MW.[9] The European market covers approximately half of European CO_2 emissions. Permits markets have already been introduced in the US, such as for lead in gasoline, SO_2 and NO_x. For greenhouse gases, a domestic emissions trading scheme has been introduced in the UK in 2002, and Denmark has created a market for CO_2 in 2000. The EU ETS draws on the US sulfur dioxide trading system for much of its inspiration, but relies much more heavily on decentralized decision making for the allocation of emissions allowances and for the monitoring and management of sources (Kruger et al. (2007, [22])).

Since 2005, the EU ETS has operated independently from the Kyoto Protocol, but it has been linked to International Emissions Trading (IET) in 2008. In this

---

[9]The sectors covered include power generation, mineral oil refineries, coke ovens, iron and steel and factories producing cement, glass, lime, brick, ceramics, pulp and paper.

framework, the 'Linking Directive' provides the recognition of Kyoto projects credits from 2008 for compliance use within the EU ETS.[10] Indeed, to provide flexibility
to the Kyoto Protocol and to reduce abatement costs, several mechanisms have been
introduced: (i) the exchange of quotas between countries, which constitutes the tradable permits market itself, (ii) the possibility to gain and exchange extra credits for
projects conducted in Joint Implementation (JI) in transition countries and that fall
under the scope of the Clean Development Mechanism (CDM),[11] and (iii) the possibility to bank unused permits during 2008–2012.

One allowance exchanged on the EU ETS corresponds to one ton of CO_2 released
in the atmosphere, and is called an European Union Allowance (EUA). 2.2 billion
allowances per year have been distributed during Phase I (2005–2007). 2.08 billion
allowances per year will be distributed during Phase II (2008–2012). With a value of
around 20 Euro per allowance, the launch of the EU ETS thus corresponds to a net
creation of wealth of around 40 billion Euro per year. As detailed above, the methodology used consisted in a free distribution of allowances during Phases I and II on
the basis of recent emissions benchmarks discussed with industrials (with minimum
auctioning). On January 2008, the European Commission has extended the scope
of the EU trading system to other sectors such as aviation and petro-chemicals by
2013, and confirmed its functioning until 2020. The allocation methodology will
shift towards auctioning during Phase III.

The EU ETS is a *compliance* market, which means that each installation of the
approximately 11,300 covered installations needs to surrender each year a number
of allowances, fixed by each Member State in its National Allocation Plan (NAP),
equal to its verified emissions [4]. In 2010, verified emissions in the EU ETS sectors
has been rising by 2.5% in 2010, from 1,869 million ton of CO_2 in 2009 to 1,915
million ton of CO_2. The amount of surplus allowances is equal to 55 million ton.
The quantity of Kyoto credits used for compliance is equal to 140 million ton, i.e.
approximatively 7% of the allowances surrendered.

Power generation contributes to more than 50% of emissions reductions in the
EU ETS [3, 4]. The main reasons are (i) the perceived ability of this sector to reduce emissions with low abatement costs (through the switching between coal and
gas or the possibility to use low-carbon technologies such as renewable energy or
carbon capture and storage), and (ii) the non-exposure of this sector to international,
non-EU competition in opposition to the main industrial sector (cement, steel, glass
production, etc.). Electricity generation from coal is cheaper than from natural gas.
However, coal is nearly twice as carbon-intensive as gas, and the introduction of
carbon costs may make gas more interesting than coal to produce electricity. Thus,
the CO_2 price path may be greatly impacted by fuel-switching decisions.

---

[10]Note that the European registry for emissions verification—the Community Independent Transaction Log (CITL)—has been connected to the Kyoto Protocol's International Transaction Log
(ITL) registry on October 16, 2008.

[11]According to the article 12 of the Kyoto Protocol, CDM projects consist in achieving greenhouse
gases emissions reduction in non-Annex B countries. After validation, the UNFCCC delivers credits that may be used by Annex B countries for use towards their compliance position.

1.3.2 Calendar

The compliance of any given installation regulated by the EU ETS follows a precise calendar. By April 30, installations need to report in their national registry their verified emissions for the previous year. The number of verified emissions needs to match the number of allowances that were initially allocated during the corresponding trading period, otherwise they need to pay a penalty.

This calendar may be summarized as follows for any compliance period starting in January of year N. On February 28 of year N, the installation receives its allocation for year N. By April 30 of year N, it needs to surrender to their national registry the allowances valid during the year $N - 1$. Therefore, from February 28 until April 30, the installation has a double accounting period, with allowances valid for year N and $N - 1$ in its registry. On May 15 of year N (at the latest), the EU Commission—which oversees national registries—publishes the verified emissions report for all installations corresponding to the trading period of the year $N - 1$. The compliance year ends on December 31 of year N. Hence, each year during February to April, compliance events occur on the European carbon market.

1.3.3 Penalties

During Phase I, if an installation does not meet its emissions target during the compliance year under consideration, the penalty is equal to 40 Euro per ton of CO_2 in excess, plus the restitution of one allowance during the next compliance period. During Phase II, this amount is equal to 100 Euro per ton, based on the same principle.

1.3.4 Market Players

As a consequence of the strong recent development of carbon markets and especially their substantial future growth potential, a vivid industry with hundreds of participating firms, asset management firms, brokers, consultants, verification agencies or other institutions has developed.

Two broad categories of actors may be distinguished on the European carbon market. First, industrials total around 11,300 installations which correspond to the current perimeter of the scheme. As recalled above, the CO_2 allowance market is heavily concentrated towards the power sector. Approximately three quarters of the yearly allocation is distributed among hundred companies. This may be seen as a typical case of market power on emission permit markets (Hintermann (2011, [18])).

Second, financial intermediaries participate to allowance trading. Their role consists in facilitating the access of industrial operators to the market, in managing part of the risks associated with emissions trading, and in providing liquidity to the market. Among financial intermediaries, brokers do not buy or sell CO_2 allowances for themselves but for their clients. They can operate through exchanges or over-the-counter. Brokers can be part of investment banks or specialised brokering agencies. On the contrary, traders can buy or sell CO_2 allowances directly to make a profit. Investment banks typically engage in that type of activity, along more traditional consulting work for their clients. Companies which receive allowances can also directly trade CO_2 allowances for profit, if they open a dedicated trading desk. EDF Trading, Total Trading or Shell Trading may be cited as examples. Finally, trading CO_2 allowances can also be seen as having a diversifying effect in portfolio management (see Chap. 5).

1.4 EUA Price Development

This section describes the main characteristics of EU ETS contracts, along with an analysis of the EU CO_2 allowance price development and descriptive statistics.

1.4.1 Structure and Main Features of EU ETS Contracts

To comply with their emissions target, installations may exchange quotas either over-the-counter, or through brokers and market places. BlueNext[12] is the market place dedicated to CO_2 allowances based in Paris. It has been created on June 24, 2005 and has become the most liquid platform for spot trading.[13] The European Climate Exchange (ECX) is the market place based in London. It has been created on April 22, 2005 and is the most liquid platform for futures and option trading.[14] Other exchanges are worth mentioning: (i) NordPool, which represents the market place common to Denmark, Finland, Sweden, Norway, and is based in Oslo; (ii) the European Energy Exchange (EEX), based in Leipzig, trading spot and derivatives products for emissions allowances rights; and (iii) the New York Mercantile Exchange (NYMEX), based in the US, which is also trading European futures and options emissions rights. The price of products exchanged on these market places are strongly correlated, which is also a feature of stock markets.

As for the main characteristics in terms of transactions, the futures contract is a deliverable contract where each member with a position open at cessation of trading for a contract month is obliged to make or take delivery of emission allowances to or from national registries. The unit of trading is one lot of 1,000 emission allowances. Each emission allowance represents an entitlement to emit one ton of carbon dioxide equivalent gas. Market participants may purchase consecutive contract months from December 2008 to December 2012. Spreads between two futures contracts may also be traded. Delivery occurs by mid-month of the expiration contract date.

---

[12]Formerly called *Powernext Carbon*.

[13]72% of the volume of spot contracts are traded on Bluenext (Reuters).

[14]96% of the volume of futures contracts are traded on ECX (Reuters).

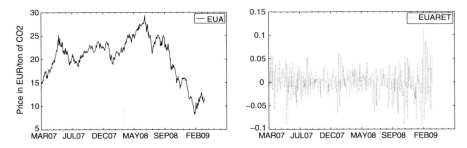

Fig. 1.1 EUA: Raw data (*left panel*) and logreturns (*right panel*) from March 09, 2007 to March 31, 2009

1.4.2 Carbon Price

In Fig. 1.1, the carbon price pictured is the rolled-over December futures contract for EUAs valid for trading under the EU ETS. It is traded in Euro/ton of CO_2 on the ECX. Phase I prices were valid during 2005–2007, while Phase II prices cover the period going from 2008 to 2012. Since Phase I carbon spot prices did not send a reliable price signal [1], we choose to work with the rolled-over December futures contract series instead.

In Fig. 1.1, it is interesting to look at the early evolution of the EUA price. Beginning at 8 Euro/ton of CO_2 on January 1, 2005, EUA prices have been rising to 25–30 Euro/ton until the release of 2005 verified emissions. On April 24, 2006 EUA prices experienced a sharp break due to the first compliance disclosures by Member States (Alberola et al. (2008, [2])). Verified emissions were about 80 million tons (or 4%) lower than the amount of allowances distributed to installations during the 2005 compliance period (Ellerman and Buchner (2008, [11])). Since the official report from the European Commission,[15] the EUA market has been sending two price signals responding to different dynamics. As the 2005 compliance confirmed the allowance market is over-supplied and the European Commission reaffirmed on October 2006 its will to enforce tighter targets during Phase II, the EUA spot price and December 2007 futures price have been declining towards zero, whereas the December 2008 futures price exhibits an increasing price pattern of 25 Euro/ton at the end of 2007.

The divorce between the EUA spot price, the futures price of delivery December 2007 on the one hand, and the futures price of delivery December 2008 on the other hand, in other words the disconnection between EUA prices of Phase I (2005–2007) and prices of Phase (2008–2012), is largely explained by the end of the allowances validity after the 2007 compliance that occurred on April 2008 due to the impossibility to transfer unused allowances to the next commitment period (on this topic, see the econometric analysis by Alberola and Chevallier (2009, [1])).

---

[15]See the EU Press Release IP/06/612. Available at http://www.europa.eu.

More recently, in Fig. 1.1, we observe that the price development features a lower bound below 10 Euro/ton of CO_2 in February 2009 (in the midst of the financial crisis), and an upper bound around 29 Euro/ton of CO_2 in May-June 2008 (consecutive to the 2007 compliance event). Analysts estimate that the total volume of transactions on the European carbon market (including exchange-based as well as over-the-counter trading) has been growing from 262 million ton in 2005 to 809 million ton in 2006, 1,455 million ton in 2007, 2,713 million ton in 2008, 5,016 million ton in 2009, and 4,414 million ton in 2010.

1.4.3 Descriptive Statistics

Box 1.1 (Normality Tests)   Let $\mu_k = \frac{1}{n}\sum_{i=1}^{n}(x_i - \overline{x})^k$ the centered moment of order k. The skewness coefficient is defined as:

$$\beta_1^{1/2} = \frac{\mu_3}{\mu_2^{3/2}} \tag{1.10}$$

with $\beta_1^{1/2} \sim N(0; \sqrt{\frac{6}{n}})$. The kurtosis coefficient is defined as:

$$\beta_2 = \frac{\mu_4}{\mu_2^2} \tag{1.11}$$

with $\beta_2 \sim N(3; \sqrt{\frac{24}{n}})$. The respective statistics can be compared to 1.96 (at the 5% level):

$$v_1 = \frac{\beta_1^{1/2} - 0}{\sqrt{\frac{6}{n}}} \tag{1.12}$$

$$v_2 = \frac{\beta_2 - 3}{\sqrt{\frac{24}{n}}} \tag{1.13}$$

The null hypothesis of normality $H_0 : v_1 = 0$ (symmetry) and $H_0 : v_2 = 0$ (normal kurtosis) is rejected if $|v_1| > 1.96$ and $|v_2| > 1.96$.

If $\beta_1^{1/2} \sim N$ and $\beta_2 \sim N$, then the Jarque-Bera (JB) statistic can be computed as follows:

$$s = \frac{n}{6}\beta_1 + \frac{n}{24}(\beta_2 - 3)^2 \tag{1.14}$$

The JB statistic is distributed as $\chi_{1-\alpha}^2$ with two degrees of freedom. If $s \geq \chi_{1-\alpha}^2(2)$, the null hypothesis of normality can be rejected at the level α.

Table 1.1 Descriptive
statistics for the carbon price
in raw (EUA) and logreturns
(EUARET) transformation

	EUA	EUARET
Mean	20.4038	−0.0004
Median	21.5200	0.0001
Maximum	29.3300	0.1136
Minimum	8.2000	−0.0943
Standard Deviation	4.4592	0.0268
Skewness	−0.7659	−0.0608
Kurtosis	3.0319	4.8680
Jarque-Bera (JB)	51.7501	77.0950
Probability (JB)	0.0001	0.0001
Observations	529	528

Source: ECX

In Table 1.1, we observe that the mean carbon price over the period is equal
to 20 Euro/ton of CO_2. These descriptive statistics also reveal that carbon prices
present negative skewness and excess kurtosis.[16] These summary statistics therefore
reveal an asymmetric and leptokurtic distribution.[17]

References

1. Alberola E, Chevallier J (2009) European carbon prices and banking restrictions: evidence
   from Phase I (2005–2007). Energy J 30:51–80
2. Alberola E, Chevallier J, Cheze B (2008) Price drivers and structural breaks in European
   carbon prices 2005-07. Energy Policy 36:787–797
3. Alberola E, Chevallier J, Cheze B (2008) The EU emissions trading scheme: the effects of
   industrial production and CO_2 emissions on European carbon prices. Int Econ 116:93–125
4. Alberola E, Chevallier J, Cheze B (2009) Emissions compliances and carbon prices under the
   EU ETS: a country specific analysis of industrial sectors. J Policy Model 31:446–462
5. Aldy JE, Stavins RN (2007) Architectures for agreement: addressing global climate change in
   the Post-Kyoto world. Cambridge University Press, Cambridge
6. Chevallier J (2010) The impact of Australian ETS news on wholesale spot electricity prices:
   an exploratory analysis. Energy Policy 38:3910–3921
7. Chevallier J (2011) Banking and borrowing in the EU ETS: a review of economic modelling,
   current provisions and prospects for future design. J Econ Surv. doi:10.1111/j.1467-6419.
   2010.00642.x
8. Chevallier J, Jouvet PA, Michel P, Rotillon G (2009) Economic consequences of permits allocation rules. Int Econ 120:77–90
9. Dales J (1968) Pollution, property and prices. Toronto University Press, Toronto
10. Dinan T, Rogers DL (2002) Distributional effects of carbon allowance trading: how govern-
    ment decisions determine winners and losers. Natl Tax J 55:199–221

---

[16]Note for a normally distributed random variable skewness is zero, and kurtosis is three.

[17]Such a fat-tailed distribution may suggest a GARCH modeling as GARCH models better accommodate excess kurtosis in the data. See the GARCH modeling framework in Chap. 3.

11. Ellerman AD, Buchner BK (2008) Over-allocation or abatement? A preliminary analysis of the EU ETS based on the 2005-06 emissions data. Environ Resour Econ 41:267–287
12. Ellerman AD, Joskow PL, Schmalensee R, Montero JP, Bailey E (2000) Markets for clean air: the US acid rain program. Cambridge University Press, Cambridge
13. Ellerman AD, Montero JP (2007) The efficiency and robustness of allowance banking in the US acid rain program. Energy J 28:47–71
14. Godby RW (2000) Market power and emissions trading: theory and laboratory results. Pacific Econ Rev 5:349–363
15. Grüll G, Taschini L (2011) Cap-and-trade properties under different hybrid scheme designs. J Environ Econ Manag 61:107–118
16. Hahn RW (1984) Market power and transferable property rights. Q J Econ 99:753–765
17. Helfand GE, Moore MR, Liu Y (2006) Testing for dynamic efficiency of the sulfur dioxide allowance market. University of Michigan working paper
18. Hintermann B (2011) Market power, permit allocation and efficiency in emission permit markets. Environ Resour Econ 49:327–349
19. Jouvet PA, Michel P, Rotillon G (2005) Optimal growth with pollution: how to use pollution permits? J Econ Dyn Control 29:1597–1609
20. Jouvet PA, Michel P, Rotillon G (2005) Equilibrium with a market of permits. Res Econ 59:148–163
21. Kling C, Rubin J (1997) Bankable permits for the control of environmental pollution. J Public Econ 64:101–115
22. Kruger J, Oates WE, Pizer WA (2007) Decentralization in the EU emissions trading scheme and lessons for global policy. Discussion paper 07-02, Resources for the Future, Washington DC, USA
23. Leiby P, Rubin J (2001) Intertemporal permit trading for the control of greenhouse gas emissions. Environ Resour Econ 19:229–256
24. Liski M, Montero JP (2005) A note on market power in an emissions permit market with banking. Environ Resour Econ 31:159–173
25. Markowitz H (1959) Portfolio selection: efficient diversification of investments. Wiley, New York
26. Montgomery WD (1972) Markets in licenses and efficient pollution control programs. J Econ Theory 5:395–418
27. Newell R, Pizer W, Zhang J (2000) Managing permit markets to stabilize prices. Environ Resour Econ 31:133–157
28. Pindyck RS (1993) The present value model of rational commodity pricing. Econ J 103:511–530
29. Rubin J (1996) A model of intertemporal emission trading, banking, and borrowing. J Environ Econ Manag 31:269–286
30. Schennach SM (2000) The economics of pollution permit banking in the context of title IV of the 1990 clean air act amendments. J Environ Econ Manag 40:189–210
31. Slade ME, Thille E (1997) Hotelling confronts CAPM: a test of the theory of exhaustible resources. Can J Econ 30:685–708
32. Sharpe WF (1964) Capital asset prices: a theory of market equilibrium under conditions of risk. J Finance 19:425–442
33. Tietenberg TH (2006) Emissions trading: principles and practice, 2nd edn. RFF Press, Washington

Chapter 2
CO$_2$ Price Fundamentals

Abstract This chapter details the main carbon price drivers related to (i) institutional decisions, (ii) energy prices and (iii) weather events. First, we review the econometric techniques based on dummy variables and structural breaks which may be useful to detect how institutional news announcements affect the price path of carbon. Second, we detail how energy prices related to fuel use (brent, natural gas, coal) and power producers' fuel-switching behavior (clean dark spread, clean spark spread, switch price) impact carbon price changes through various regression models. Third, we detail the economic rationale behind the influence of extreme weather events on carbon prices, based on unanticipated temperatures changes. The Appendix presents an extension of the econometric framework on the basis of multivariate GARCH modelling between energy and CO$_2$ prices.

2.1 Institutional Decisions

The price path of CO$_2$ allowances under the EU ETS is deeply affected by institutional decisions. Indeed, the EU ETS is an environmental market, which has been created by the Directive 2003/87/EC. According to this environmental regulation tool, the EU Commission and more particularly the Directorate-General for Climate Action (DG CLIMA) can intervene to amend the functioning of the scheme. As market participants fully react to news arrival, any policy intervention from the regulator has an immediate impact on the carbon price (Mansanet-Bataller and Pardo ([41], 2009), Daskalakis and Markellos ([20], 2009), Conrad et al. ([17], 2011)).

2.1.1 Dummy Variables

The influence of institutional decisions is classically captured in time-series econometrics by using dummy variables. We explain the main logic behind using this econometric tool, along with a hands-on example.

2.1.1.1 Basic Principle

Dummy variables are useful to capture the presence or the absence of a given phenomenon. They can be introduced in the multiple regression model when one wants to take into account, among other explanatory variables, the effect of a binary explanatory factor. These binary variables are composed of zeros throughout the period (chronologically ordered in days) in the absence of the phenomenon under consideration. When one institutional event is expected to affect the carbon price, it is coded as one in the dummy variable. For instance, a news announcement by the EU Commission concerning changes in the allocation methodology for the EU ETS is supposed to be immediately integrated by market agents and it will be reflected in the current carbon price. On that specific day, we will set the dummy variable to be equal to one.

Next, by regressing carbon price changes on the dummy variable, we evaluate the significant impact of such institutional events on the price path of carbon at usual statistical levels. As mentioned earlier, the main interest behind dummy variables lies in their ability to be added to other more traditional explanatory variables. It is noteworthy to remark that the introduction of dummy variables modifies neither the estimation methods, nor the statistical tests, of the multiple regression model.

2.1.1.2 A Practical Example

The following example can be replicated by the interested reader by using the R software.[1] The dataset used in this example can be downloaded at: http://sites.google.com/site/jpchevallier/publications/books/springer/data_chapter2.csv.

To illustrate how to use dummy variables, we wish to evaluate the impact of news about Phase II and Phase III National Allocation Plans (NAPs) by the European Commission on the price of carbon. To do so, we have identified the news announcements summarized in Table 2.1. The dummy variables refer to news information disclosure concerning NAPs Phase II (*NAPsPhaseII*), and Phase III (*NAPsPhaseIII*). The main sources for the news events recorded may be cited as being: the UNFCCC, the European Commission, the European Parliament, the European Economic and Social Committee, ECX, EEX, Bluenext, ICE, and Point Carbon.

The period under consideration goes from March 9, 2007 to March 31, 2009, i.e. it covers a sample of 529 daily observations. During the days when no news event falling in either of the two categories was recorded, the dummy variables take the value of zero as explained in Sect. 2.1.1.1.

Box 2.1 (Stationarity and Unit Root Tests) Before estimating any time series model, it is necessary to test for the stationarity of the dependent variable (and independent variables as well in ARMAX models). Dickey-Fuller (1979, [24]) test the nullity of the coefficient α in the following regression:

$$\Delta x_t = x_{t+1} - x_t = \alpha x_t + \beta + e_t \qquad (2.1)$$

---

[1]R is a free software environment for statistical computing and graphics. It can be downloaded at: http://www.r-project.org/. See [46] for an introduction.

Table 2.1 Dummy Variables

Date	NAPs Phase II	NAPs Phase III
26/03/2007	1	0
26/03/2007	1	0
02/04/2007	1	0
04/05/2007	1	0
15/05/2007	1	0
04/06/2007	1	0
13/07/2007	1	0
18/07/2007	1	0
31/08/2007	1	0
22/10/2007	1	0
26/10/2007	1	0
26/10/2007	1	0
26/10/2007	1	0
07/12/2007	1	0
23/01/2008	0	1
28/02/2008	0	1
03/03/2008	0	1
05/06/2008	0	1
06/06/2008	0	1
11/06/2008	0	1
09/07/2008	0	1
07/10/2008	0	1
08/10/2008	0	1
15/10/2008	0	1
20/10/2008	0	1
04/12/2008	0	1
17/12/2008	0	1
17/12/2008	0	1
17/12/2008	0	1

Source: UNFCCC, European Commission, European Parliament, European Economic and Social Committee, ECX, EEX, Bluenext, ICE, Point Carbon

- if α is significantly negative, then we say that the process x_t has no unit root, or that it is stationary, inducing a mean-reverting behavior for the prices;
- if α is not significantly different from 0, then we say that the process x_t 'has a unit root', inducing a random walk behavior for the prices.

In practice, the Augmented-Dickey-Fuller (1981, ADF, [25]) or Phillips-Peron (1988, PP, [44]) tests are used rather than Dickey-Fuller. These tests are based on the same principle but correct for potential serial autocorrelation and time trend in Δx_t through a more complicated regression:

$$\Delta x_t = \sum_{i=1}^{L} \beta_i \Delta x_{t-i} + \alpha x_t + \beta_1 t + \beta_2 + e_t \qquad (2.2)$$

The ADF test tests the null hypothesis H_0 that $\alpha = 0$ (the alternative hypothesis H_1 being that $\alpha < 0$) by computing the Ordinary Least Squares (OLS) estimate of α in the previous equation and its t-statistics $\hat{t}$; then, the statistics of the test is the t-statistics $\hat{t}_\alpha$ of coefficient α, which follows under H_0 a known law (studied by Fuller and here denoted *Ful*). The test computes the p-value p, which is the probability of $Ful \leq \hat{t}$ under H_0. If $p < 0.05$, H_0 can be safely rejected and H_1 accepted: we conclude that the series 'x_t has no unit root'. Extensions of these stationarity tests were also developed by Kwiatkowski, Phillips, Schmidt, and Shin (1992, KPSS, [39]). This task can easily be performed in R, as shown below:

```
1 library(tseries)
2 adf.test(eua)
3 PP.test(eua)
4 kpss.test(eua)
```

Line 1 loads the relevant library. Line 2 performs the ADF test on the properly transformed EUA time series. Lines 3 and 4 compute, respectively, the PP and KPSS tests. The user can also choose between various lags (command `lags`) and deterministic (command `type="trend"`) or stochastic (command `type="drift"`) specifications for the tests (see [8] for a review).

To capture the influence of these institutional news events on the price of carbon, we run the following regression:

$$y_t = \alpha + \phi y_{t-1} + \beta NAPsPhaseII_t + \theta NAPsPhaseIII_t + \varepsilon_t \qquad (2.3)$$

with y_t the dependent carbon price variable (taken in log first-difference), α the constant term, ϕ the parameter capturing the influence of the AR(1) process, β the parameter capturing the influence of the dummy variable *NAPsPhaseII*, θ the parameter capturing the influence of the dummy variable *NAPsPhaseIII*, and ε_t the error term.

Box 2.2 (ARIMA(p, d, q) Modeling of Time Series) According to Wold's theorem, a non-stationary and stochastic time series can be modeled as an AutoRegressive Integrated Moving Average (ARIMA) process:

$$\Phi(L)(1 - L)^d X_t = \Theta(L)\varepsilon_t \qquad (2.4)$$

with L the lag operator, p the AR order, q the MA order, and d the integration order. The time series needs to be transformed d times to exhibit stationarity properties (see Hamilton (1996, [33] for a detailed coverage).

Table 2.2   R Code: Linear regression with dummy variables

```
1   path<-"C:/"
2   setwd(path)
3   library(dynlm)
4   data=read.csv("data_chapter2.csv",sep=",")
5   attach(data)
6   eua=data[,2]
7   napII=data[,3]
8   napIII=data[,4]
9   model<-dynlm(eua~L(eua)+napII+napIII)
10  summary(model)
11  layout(matrix(1:4,2,2))
12  plot(model)
13  acf(residuals(model))
14  acf(residuals(model),type='partial')
```

The ARIMA modeling of a given time series can be achieved in R with the commands:

```
1 library(stats)
2 model<-arima(eua,order=c(1,0,0))
```

Line 2 features the AR(1) part of the data generating process, which can contain three components (p, d, q). p is the AR order. d is the degree of differencing. q is the MA order (see [33]). Note that external regressors can optionally be added with the argument `xreg`.

To estimate the dynamic linear model in Eq. (2.3) with time series regression relationships, the interested reader may run the computer code contained in Table 2.2 under the R software.

Lines 1–2 specify the user's working directory. Line 3 loads the R library 'dynlm', which needs to be pre-installed by the user. Lines 4 to 8 load the data. Line 9 estimates the dynamic linear model in Eq. (2.3). Line 10 produces the regression results shown in Table 2.3.

We can see that the coefficient estimate for the AR(1) process is highly significant (at the 1% level). Other coefficient estimates for the dummy variables *NAPsPhaseII* and *NAPsPhaseIII* are not statistically significant. In this illustrative example, we fail to detect the effects of news announcements about allocation methodologies on the price of carbon.

Looking at the residuals estimated, the output gives us a superficial view of the distribution of the residuals, which may be used as a quick check of the distributional assumptions [22]. Note that the average of the residuals is zero by definition, so the median should not be far from zero, and the minimum and maximum should roughly be equal in absolute value. In our setting, the first and third quartiles are remarkably close to zero.

Table 2.3 R Output: Linear regression with dummy variables

```
Call:
dynlm(formula = eua ~ L(eua) + napII + napIII)

Residuals:
      Min          1Q      Median          3Q         Max
-1.289e-17  -5.985e-19  -1.492e-19   4.469e-19   2.938e-17

Coefficients:
              Estimate Std. Error    t value Pr(>|t|)
(Intercept)  0.000e+00  9.940e-20  0.000e+00    1.000
L(eua)       1.000e+00  2.684e-18  3.725e+17   <2e-16 ***
napII       -4.338e-19  6.519e-19 -6.650e-01    0.506
napIII       5.498e-19  6.271e-19  8.770e-01    0.381
---
Signif. codes:  0 '***' 0.001 '**' 0.01 '*' 0.05 '.' 0.1 ' ' 1

Residual standard error: 2.231e-18 on 525 degrees of freedom
Multiple R-squared:       1,      Adjusted R-squared:       1
F-statistic: 4.637e+34 on 3 and 525 DF,  p-value: < 2.2e-16
```

Next, the residual standard error is the residual variation, i.e. an expression of the variation of the observations around the regression line, estimating the variance of the model parameters. Besides the R-squared and Adjusted R-squared statistics, the F-statistics is an F test for the hypothesis that the regression coefficient is zero. These latter diagnostic tests are not really interesting in this practical example. They will be more useful later in this chapter as we introduce more explanatory variables.

Finally, lines 11 to 14 show several residual plots. Let us describe the different panels in Fig. 2.1. The upper left panel provides a graph of residuals versus fitted values, perhaps the most familiar of all diagnostic plots for residual analysis [38]. It is useful for checking the assumption that $E(\varepsilon|X) = 0$, as the regressions and confidence intervals used in Eq. (2.3) are based on the assumption of i.i.d. normal errors. In our setting, we can observe some clustering and irregular variations in the residuals, which suggests that the model is mis-specified. The bottom left panel is a Quantile-Quantile plot for normality, whose curvature (or rather its lack thereof) shows whether the residuals are more or less conform with normality. The insight behind a QQ-plot is that the residuals can be assessed for normality by comparing them to "ideal" normal observations. The upper and bottom right panels are a scale-location plot and a plot of standardized residuals against leverages, respectively. They are useful for checking the assumption that errors are identically distributed, and to display combinations of standardized residuals, leverage, and Cook's distance.[2]

---

[2]For further details on how to interpret regression diagnostics for linear models, see [38].

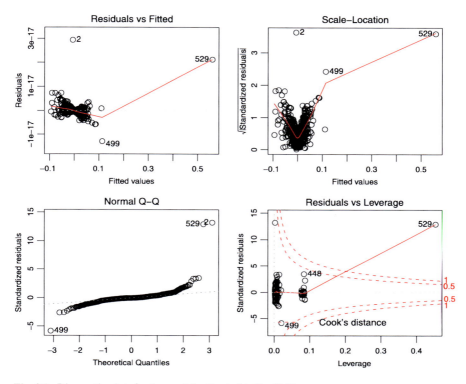

Fig. 2.1 Diagnostic plots for the model estimated in Eq. (2.3)

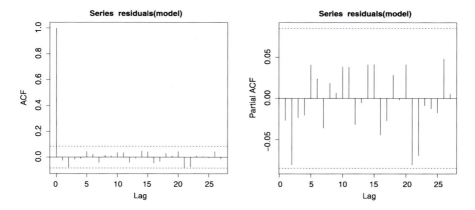

Fig. 2.2 Autocorrelation and partial-autocorrelation functions for the residuals from the model estimated in Eq. (2.3)

Figure 2.2 shows the autocorrelation (ACF) and partial-autocorrelation (PACF) functions of the residuals. The broken horizontal lines on the plots correspond to

95-percent confidence limits. The general pattern of the autocorrelation and partial autocorrelation functions—one positive spike in the former; sinusoidal decay in the latter—is indicative of an AR(1) process.

Box 2.3 (The Ljung-Box-Pierce Residual Autocorrelation Test)   Under the null hypothesis of no autocorrelation in the residuals, noted $\hat{\varepsilon}_t = \frac{\hat{\Phi}(L)}{\hat{\Theta}(L)} X_t$:

$$\hat{\rho}_{1_{(\hat{\varepsilon}_t)}} = \hat{\rho}_{2_{(\hat{\varepsilon}_t)}} = \cdots = \hat{\rho}_{k_{(\hat{\varepsilon}_t)}} = 0 \tag{2.5}$$

the Box-Pierce 'portmanteau' test statistic writes:

$$BP(K) = T \sum_{k=1}^{K} \hat{\rho}^2_{k_{(\hat{\varepsilon}_t)}} \tag{2.6}$$

with T the number of observations, $\hat{\rho}_{k_{(\hat{\varepsilon}_t)}}$ the autocorrelation coefficient of order k of the estimated residuals, and K the maximum number of lags. Under the assumption that X_t is an i.i.d. sequence with certain moment conditions, $BP(K)$ is asymptotically a χ^2 random variable with $(K - p - q)$ degrees of freedom (see Hamilton (1996, [33]) for further details).

Ljung-Box have modified this statistic to increase the power of the test in finite samples:

$$LB(K) = T(T + 2) \sum_{k=1}^{K} \frac{\hat{\rho}^2_{k_{(\hat{\varepsilon}_t)}}}{T - k} \tag{2.7}$$

Similarly, $LB(K)$ is asymptotically a χ^2 random variable with $(K - p - q)$ degrees of freedom.

To test formally for the presence of autocorrelation in the residuals, the Ljung-Box-Pierce test can be performed in R:

```
1 library(tseries)
2 Box.test(eua)
```

This code computes the Box-Pierce or Ljung-Box test statistic for examining the null hypothesis of independence in a given time series (see [34]).

Next, we explore another methodology to detect the effects of institutional decisions on the price path of carbon, based on structural break tests.

2.1.2 Structural Breaks

In what follows, we explain first the rationale behind relating institutional decisions to structural breaks. Second, we provide an empirical application to the EUA time series.

Table 2.4 R Code: Bai-Perron Structural Break Test

```
1   path<-"C:/"
2   setwd(path)
3   library(strucchange)
4   data=read.csv("data_chapter2.csv",sep=",")
5   attach(data)
6   eua=data[,2]
7   bp.ri <- breakpoints(eua ~ 1, h = 30)
8   summary(bp.ri)
9   plot(bp.ri)
10  coef(bp.ri, breaks = 3)
11  vcov(bp.ri, breaks = 3, vcov = kernHAC)
12  confint(bp.ri, breaks = 3, vcov = kernHAC)
```

2.1.2.1 Structure of the Test

The influence of institutional decisions on the price path of carbon may also be detected by means of structural break tests. Formally, the concept of structural break may be defined as changes in the underlying structural relationship within an economy. More intuitively, it can be said that structural breaks occur in a time series when the structure of the data generating mechanism underlying a set of observations changes. Therefore, our econometric approach consists here in detecting when shocks occur in the carbon price series, and to attempt to relate such shocks to institutional news announcements by the EU Commission.

Bai and Perron have developed very popular techniques to estimate and test linear models with multiple structural changes [5–7]. By borrowing notations from [47], we briefly recall below the multiple linear regression model with m breaks (or, equivalently, $m + 1$ regimes):

$$y_t = x_t^\top \delta_j + u_t, \quad t = T_{j-1} + 1, \ldots, T_j, \ j = 1, \ldots, m + 1 \tag{2.8}$$

with T the sample size, $T_0 = 0$ and $T_{m+1} = T$. The goal of the analysis is to determine the number and location of the breakpoints $T_j, j = 1, \ldots, m$. A search over m is conducted, for $m \leq m^*$, where m^* is fixed by the researcher. The minimum number of observations per segment, h, can also be exogenously set by the user.[3]

2.1.2.2 Empirical Application

Let us apply this analysis to carbon price data by using the R code contained in Table 2.4.

As previously, lines 1 to 6 load the relevant libraries (pre-installed by the user) and data. Line 7 specifies a linear model in which the dependent variable *EUA* is

---

[3]Note this may be considered as a bandwidth or trimming parameter.

Fig. 2.3 EUA: RSS and BIC
for models with up to sixteen
breaks

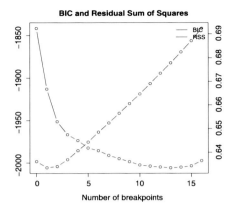

modeled by just an intercept. Line 8 produces the estimation output (not reproduced here because of its length), which comprises a triangular matrix containing the residual sums of squares (RSS), the extraction of breakpoints, corresponding information criteria, coefficient estimates and confidence intervals as described by Bai and Perron [5–7]. Line 9 plots all models with $m = 1, \ldots, 16$ breaks along with the corresponding values of the Bayesian Information Criterion (BIC) and the RSS.

As shown in Fig. 2.3, the BIC selects two breaks $m = 2$. However, because information criteria are often downward-biased, Bai and Perron argue in favor of the presence of an additional break. That is why we proceed with the estimation of a three-break model in lines 10–11. The output (not shown here) returns coefficient estimates with standard errors utilizing a kernel heteroskedasticity and autocorrelation consistent (HAC) estimator.[4] Line 12 computes the confidence intervals for the three breakpoints.[5]

In Fig. 2.4, we have reproduced the time series of EUA along with the breakpoints and corresponding confidence intervals (at the default 95% level) coded by the observation numbers and by the dates on the underlying time scale. The estimated breakpoints are:

1. May 28, 2007
2. December 30, 2008
3. February 11, 2009

There are considerable differences in the confidence intervals corresponding to the break dates. In this example, it emerges that the intervals for the second and

---

[4]With a quadratic spectral kernel, prewhitening using a VAR(1) model and an AR(1) approximation for the automatic bandwidth selection (see [47]).

[5]The confidence intervals are derived from the distribution of the argmax functional of a process composed of two independent Brownian motions with different linear drifts and scales (see [5] for further details on this nonstandard distribution). This cumulative distribution function depends on three parameters which are associated with ratios of quadratic forms in the magnitude of the shifts and weighting matrices defined as segment-wise covariance matrices [47].

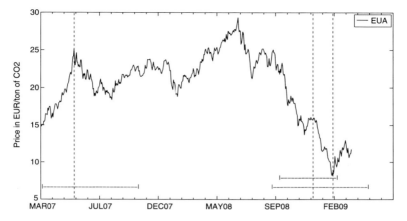

Fig. 2.4 EUA with three breaks (*vertical lines*) and 95% confidence intervals (*horizontal lines*)

third breaks overlap. Hence, there appears to be considerable uncertainty as to the location of the second break. Given that information criteria such as the BIC select a model with two breaks and that the decrease of the RSS when passing from a two break to a three break model is small (see Fig. 2.3), these results may be summarized as pointing to two competing models: $m = 2$ or $m = 3$.

Visual inspection hints at the presence of a strong adjustment period during December 2008–February 2009, which corresponds to the impact of the financial crisis on the carbon market [14]. In that case, it may be more appropriate to run separate regressions in sub-samples (before and after the crisis) in order to identify carbon price drivers. Therefore, structural break tests may also be used to delimit full- and sub-samples estimates (see [2] for another application).

The first break is somewhat of smaller magnitude than the other two, and may be related to the 2006 yearly compliance event [15]. Therefore, the first breakpoint may be directly related to the effects of institutional news by the European Commission concerning the publication of yearly verified emissions.

Overall, we have seen how to detect structural breaks in the time series of EUAs, which may then be related to institutional news announcements. In the next section, we extend our analysis to another kind of carbon price drivers: energy prices.

2.2 Energy Prices

The task faced by the econometrician when investigating the empirical properties of carbon prices consists in finding the appropriate market fundamentals. On the supply side, the amount of CO_2 allowances distributed each year according to NAPs are known in advance by all market participants. Hence, there is no uncertainty concerning the number of allowances flowing on the market. On the demand side, carbon prices are impacted by energy prices because it relates to the producing process of the utilities regulated by the EU ETS. As companies pollute (i.e. they emit tons of

CO$_2$ in the atmosphere), they face essentially two choices: either compensate for the associated pollution (by surrendering allowances from their own registry), or purchase/sell allowances on the market depending on the excess/lack of pollution compared to their annual endowment (and thereby engaging in permits trading). The amount of permits that a given company needs in order to balance its actual CO$_2$ emissions depends on the type of fuel burned. Coal for instance is far more CO$_2$ intensive than natural gas. Companies need also to pay attention to the relative variation of energy prices: in case of a rise in the price of oil for instance, the price of many manufactured products (using oil as an input to production or to produce energy) will similarly increase. That is why the price of carbon, and the demand for CO$_2$ allowances, are linked to the variation of energy prices. Finally, we need to keep in mind that power producers are key players on the carbon market, as they received globally around 50% of allocation each year. Therefore, their fuel-switching behavior (producing one unit of electricity based on coal-fired or gas-fired power-plants, and switching between inputs as their relative price vary) will be a central factor to explain the variation of the carbon price in this chapter.

In what follows, we briefly review the existing literature. Then, we consider several regression models by including oil, natural gas, coal and electricity variables, and explain how they affect the carbon price throughout our econometric analysis.

2.2.1 Literature Review

Following the pioneering work by Christiansen et al. ([16], 2005) and Kanen ([36], 2006), Ellerman and Buchner [26, 27], and Convery and Redmond ([19], 2007) produced the first literature reviews on the carbon price development in their respective articles. This work was further elaborated by Convery ([18], 2009) and Ellerman, Convery and De Perthuis ([28], 2010).

Based on economic analysis (essentially demand and supply fundamentals), Christiansen et al. ([16], 2005) have identified the following factors as being the price determinants in the EU ETS: policy and regulatory issues; market fundamentals, including the emissions-to-cap ratio, the role of fuel-switching, weather and production levels. Mansanet-Bataller et al. ([42], 2007) and Alberola et al. ([2], 2008) were the first analyses to uncover econometrically the relations between energy markets and the CO$_2$ price. Based on Phase I spot and futures data, the former group of authors establishes that carbon prices in the EU ETS are linked to fossil fuel (e.g., oil, gas, coal) use. By using an extended dataset, the latter group of authors emphasizes that the nature of this relationship between energy and carbon prices varies depending on the period under consideration, and the major influence of institutional events. In addition, Keppler and Mansanet-Bataller ([37], 2010) have studied the causalities between CO$_2$ and electricity variables (such as clean dark and clean spark spreads, and the switch price) during the first phases of the EU ETS.

Concerning the econometric modelling of the EUA time series, Benz and Trück ([9], 2009) have detailed the modeling of the price dynamics for CO$_2$ emission allowances based on GARCH models. Besides Paolella and Taschini ([43], 2008) have shown how to model the properties of CO_2 spot returns during this very specific period of the end of Phase I. On this topic, it is also worth highlighting the important theoretical contribution by Daskalakis et al. ([21], 2009) under the form of an explicit modeling of the effects of banning banking on pricing CO_2 futures prices during the first phase, and the empirical investigation of these effects by Alberola and Chevallier ([1], 2009).

Bunn and Fezzi ([12], 2009) further quantify the mutual interactions of electricity, gas and carbon prices in the UK. By using a structural, co-integrated vector error-correction model, they derive the dynamic pass-through of carbon prices into electricity prices, and the response of electricity and carbon prices to shocks in the gas price.

Finally, it is worth highlighting the work by Hintermann ([35], 2010), who derives a structural model of the allowance price under the assumption of efficient markets and examines the extent to which this variation in price can be explained by marginal abatement costs. Last but not least, macroeconomic fundamentals of carbon prices have been studied by Alberola et al. ([3], 2008, [4], 2009), as it will be more detailed in Chap. 3.

2.2.2 Oil, Natural Gas and Coal

This section investigates the impact of fuel use (brent, natural gas, coal) on carbon prices. We aim at understanding how carbon prices are impacted by the relative variation of fuel prices as an input to energy production.

2.2.2.1 Recent Evolution

The brent price (expressed in Euro/BBL) is the brent crude futures Month Ahead price negotiated on ICE (Fig. 2.5). The brent is a North Sea deposit: its oil is representative of the crudes produced in this region. Therefore, it provides the best characteristics to match other energy variables traded in continental Europe, which enter in the determination of the carbon price.

The natural gas price used (expressed in Euro/Therm) is the futures Month Ahead natural gas price negotiated on Zeebrugge Hub (Fig. 2.6). It is the most liquid gas trading market in Europe, and has a major influence on the price that European consumers pay for their gas. As such, the Zeebrugge price represents the best proxy of the European gas market price determined close to end-users.

The coal price is the Antwerp/Rotterdam/Amsterdam (ARA) coal futures Month Ahead price (see Fig. 2.7), which is the major imported coal in northwest Europe. The ARA coal is expressed in Euro/ton.

Before proceeding with our regression analysis, it is important to check the matrix of cross-correlations between the carbon price and energy variables (transformed to logreturns). Table 2.5 reveals that no cross-correlation is above 0.6 in absolute value. This step is important in order to detect multi-collinearities that could potentially bias the result estimates from the regression. If multi-collinearities are

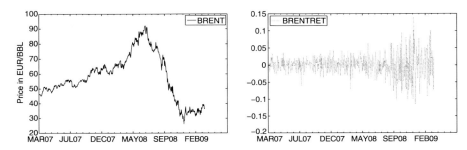

Fig. 2.5 Brent: Raw data (*left panel*) and logreturns (*right panel*) from March 09, 2007 to March 31, 2009

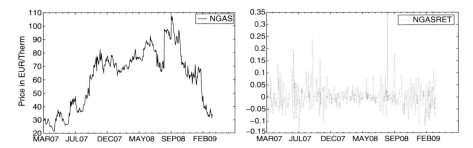

Fig. 2.6 Natural Gas: Raw data (*left panel*) and logreturns (*right panel*) from March 09, 2007 to March 31, 2009

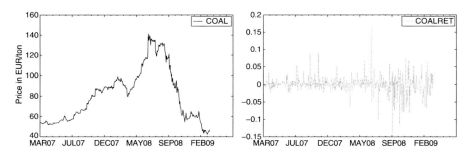

Fig. 2.7 Coal: Raw data (*left panel*) and logreturns (*right panel*) from March 09, 2007 to March 31, 2009

Table 2.5 Cross-correlations between carbon (EUA) and energy (BRENT, NGAS, COAL) variables in logreturns transformation

	EUARET	BRENTRET	NGASRET	COALRET
EUARET	1			
BRENTRET	0.2207	1		
NGASRET	0.1959	0.0978	1	
COALRET	−0.1215	−0.0792	0.0321	1

suspected, the econometrician needs to resort to more formal tests, such as the variance inflation ratio (see [32] for an extensive coverage of this topic).

Box 2.4 (Granger Causality Test)   In conjunction with the analysis of the matrix of cross-correlations, the econometrician may resort to Granger causality tests as well. These tests allow to infer causality 'in the Granger sense' between a set of dependent and independent variables selected by the user, and may be useful in econometric modelling prior to the regression analysis.

The main insight behind Granger causality testing for two variables X and Y is to evaluate whether the past values of X are useful for predicting Y once Y's history has been modeled. The null hypothesis is that the past p values of X do not help in predicting the value of Y.

The test is implemented by regressing Y on p past values of Y and p past values of X. An F-test is then used to determine whether the coefficients of the past values of X are jointly zero. Under the null hypothesis that X does not cause Y, the non-constrained model (NC) writes $Y_t = \alpha + \beta_1 Y_{t-1} + \beta_2 X_{t-1} + \varepsilon_t$. The constrained model (C) writes $Y_t = \alpha + \beta_1 Y_{t-1} + \varepsilon_t$. The Fisher test-statistic is computed as $F^* = \frac{(SSR_C - SSR_{NC})/c}{SSR_{NC}/(n-k-1)}$ with SSR the sum of squared residuals, c the number of restrictions (i.e. the number of coefficients whose nullity are tested), and k the number of variables. If F^* is superior to the critical value found in the Fisher statistical law with $(1, n-2)$ degrees of freedom, then the null hypothesis (causality from X to Y) is rejected.

The code below illustrates how to perform Granger causality testing in R:

```
1 library(MSBVAR)
2 granger.test(eua, p=2)
```

where p is the past value (or lag) set by the user. The results from linear Granger causality tests are known to be sensitive to the choice of the lag. Therefore, the user may experiment with various lags in order to test the robustness of the conclusions obtained.

2.2.2.2 Regression Analysis

To illustrate our analysis on the expected impacts of energy variables on the carbon price, we run the following regression:

$$y_t = \alpha + \beta_1 Brent_t + \beta_2 Ngas_t + \beta_3 Coal_t + \varepsilon_t \tag{2.9}$$

with *EUARET* as the dependent variable, {*BRENTRET, NGASRET, COALRET*} as independent variables, α as the constant, and ε_t as the error term.[6] This regression can be computed with R based on Table 2.6.

---

[6]Note that we do not consider here econometric models where the variance of the processes is not constant, which will be dealt with in the next chapter.

Table 2.6 R Code: Linear Regression Model with Energy Variables

```
1   path<-"C:/"
2   setwd(path)
3   library(stats)
4   data=read.csv("data_chapter2.csv",sep=",")
5   attach(data)
6   eua=data[,2]
7   brent=data[,5]
8   gas=data[,6]
9   coal=data[,7]
10  modelenergy<-lm(eua~brent+gas+coal)
11  summary(modelenergy)
12  layout(matrix(1:4,2,2))
13  plot(modelenergy)
14  acf(residuals(modelenergy))
15  acf(residuals(modelenergy),type='partial')
```

Table 2.7 R Output: Linear Regression Model with Energy Variables

```
Call:
lm(formula = eua ~ brent + gas + coal)

Residuals:
      Min        1Q     Median       3Q        Max
-0.08979  -0.01512  -0.00064   0.01309    0.55154

Coefficients:
              Estimate Std. Error t value Pr(>|t|)
(Intercept)   0.000606   0.001561   0.388   0.6980
brent         0.098255   0.052959   1.855   0.0641 .
gas           0.068477   0.035243   1.943   0.0526 .
coal         -0.134923   0.067990  -1.984   0.0477 *
---
Signif. codes:  0 `***' 0.001 `**' 0.01 `*' 0.05 `.' 0.1 ` ' 1

Residual standard error: 0.03589 on 525 degrees of freedom
Multiple R-squared: 0.02296,    Adjusted R-squared: 0.01738
F-statistic: 4.113 on 3 and 525 DF,  p-value: 0.006704
```

This code returns the output reproduced in Table 2.7.

We observe that the three energy variables considered as significant in explaining carbon price changes at usual statistical levels.[7] The coefficient estimates for the brent and natural gas variables is positive and significant at the 10% level. The coefficient estimate for the coal variable is *negative*—as expected due to its carbon-

---

[7]The user may test other specifications, for instance by introducing various lags for the dependent and independent variables.

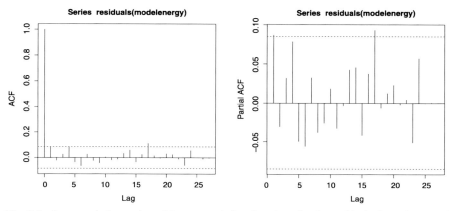

Fig. 2.8 Autocorrelation and partial-autocorrelation functions for the residuals from the model estimated in Eq. (2.9)

intensive content—and significant at the 5% level. Therefore, we are able to verify in this example that the use of fuels such as brent and natural gas has a positive impact on the carbon price. Higher demand for energy will translate into more fuel use, higher prices for brent and natural gas, and increased demand for CO_2 allowances. Conversely, the use of coal has a negative impact on the carbon price. As the more carbon-intensive input to produce energy, when the price of coal rises, agents need to reduce their consumption of coal, which leads to decreased demand for CO_2 allowances. The Adjusted R-Squared is low (0.0173), which means that the explanatory power of this regression can be improved by adding other relevant independent variables, as it will be detailed later in this chapter. The p-value of the F-test statistic (0.0067) shows that the joint significance of the results is accepted at the 1% level.

Lines 12 to 15 allow to check the residuals of the regression. We only reproduce the ACF and PACF below in Fig. 2.8. We cannot detect any systematic pattern in the residuals, which stay safely within the confidence intervals, and hence we conclude that the errors are not autocorrelated for this regression.

Next, let us explore the relationship between the carbon price and electricity variables.

2.2.3 Electricity Variables

Depending on the share of coal and gas in their electricity mix, power producers may benefit from the opportunity to switch from coal to gas depending on the price of the carbon price in order to minimize the electricity production cost.[8] The opportunity to switch provides an interesting solution for power generators to reduce their

---

[8]Note we consider here the switching between coal and gas to produce one unit of electricity. Oil is mostly used to meet high peak demand, especially in winter. Thus, oil is unlikely to interfere with coal and gas switching in the European system. Besides, gas and coal are the cheapest alternatives, while oil has encountered a high level of volatility during the recent period.

emissions at low cost. In order to benefit from these cheap abatement opportunities, electricity generators need to compare the CO$_2$ price and marginal abatement costs. It may be cheaper to switch for a less carbon-intensive fuel (such as gas) instead of using coal and buying permits. That is why, in this section, we describe the basic principle behind fuel-switching for power producers.

2.2.3.1 Fuel-Switching Behavior

In the power sector, electricity prices are determined by the marginal generation technology. Different generation units are ranked by marginal costs from the cheapest technology to the most expensive one.[9] The introduction of a carbon cost through tradable permits changes the carbon intensity of power generation technologies, and modifies the competitiveness of power plants. Through the introduction of high carbon costs, gas-fired power plants—which are more expensive but less CO$_2$-intensive than coal-fired power plants—may become more profitable than coal-fired power plants. The price of carbon that makes the two technologies equally attractive is called the 'switching point'.[10]

Basic Principle Without carbon costs, the marginal cost of producing electricity is given by the ratio between fuel costs and the efficiency of the plant. It may be calculated by using the plants efficiencies [23]:

$$MC = \frac{FC}{\eta} \tag{2.10}$$

with MC the marginal cost, FC the fuel cost, and η the efficiency of the plant. With carbon costs, the marginal cost for each plant contains the emission factor which depends on the fuel burnt:

$$MC = \frac{FC}{\eta} + \frac{EF}{\eta} \times EC \tag{2.11}$$

with EF the emission factor, and EC the emission cost. The switching point between a given coal-fired plant and a given gas-fired power plant may be computed as the emission cost which leads to $MC_{ngas} = MC_{coal}$. It represents the carbon cost that leads to a switch between two plants in the merit order. This price depends on each plant's fuel cost, efficiency and emission factors:

$$EC_{switch} = \frac{\eta_{coal} \times FC_{ngas} - \eta_{ngas} \times FC_{coal}}{\eta_{ngas} \times EF_{coal} - \eta_{coal} \times EF_{ngas}} \tag{2.12}$$

If the carbon price is lower than the switching point, electricity generation through coal is more profitable than gas.[11]

---

[9]This ranking is called 'merit order', and depends on several parameters such as fuel prices, plant efficiencies and carbon intensity.

[10]We do not include in the analysis the capital and operations and maintenance costs of coal versus natural gas. Hence, we aim at capturing the *short-term* effect of introducing emissions trading on power producers' fuel-switching behaviour. A longer term analysis requires an estimation of fuel substitution with additional costs associated to retro-fit.

[11]Also note that the potential for fuel-switching is very dependent on the load. Indeed, at full load, during winter peak-hours for instance, all the plants are running so that no opportunity to

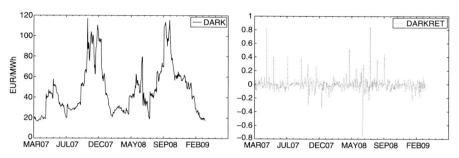

Fig. 2.9 Clean Dark Spread: Raw data (*left panel*) and logreturns (*right panel*) from March 09, 2007 to March 31, 2009

Clean Dark Spread, Clean Spark Spread and Switch Price   Power operators pay close attention to the Dark and Spark spreads, as well as to the difference between them. The Dark spread represents the theoretical profit that a coal-fired power plant makes from selling a unit of electricity, having purchased the fuel required to produce that unit of electricity. The Spark spread refers to the equivalent for natural gas-fired power plants. With the introduction of carbon costs, the Dark and Spark spreads need to be corrected by EUA prices, and thus become respectively the Clean Dark and Clean Spark spreads. The Switch price represents the threshold for the carbon price *above* which it becomes profitable for an electric power producer to switch from coal to natural gas, and *below* which it is beneficial to switch from natural gas to coal. This switching price is more sensitive to natural gas price changes than to coal prices changes [36]. Together, these three profitability indicators are used to determine the preferred fuel used for power generation. Now, let us describe in more details these electricity variables selected to study the effect of power producers' fuel-switching behavior on the carbon price.

Clean Dark Spread   First, the Clean Dark Spread (*clean dark*, expressed in Euro/MWh) represents the difference between the price of electricity at peak hours (*elec*) and the price of coal (*coal*) used to generate that electricity, corrected for the energy output of the coal plant and the costs of CO_2 (p_t):

$$clean\ dark = elec - \left(coal \times \frac{1}{\rho_{coal}} + p_t \times EF_{coal} \right) \qquad (2.13)$$

with ρ_{coal} the net thermal efficiency of a conventional coal-fired plant (i.e. 35% according to Reuters), and EF_{coal} the CO_2 emissions factor of a conventional coal-fired power plant (i.e. 0.95 tCO_2/MWh according to Reuters). The Clean Dark Spread is shown in Fig. 2.9.

---

switch exists. The switch can only occur if coal-plants are running while some gas-fired plants are available to replace them. The best opportunities occur when the load is relatively low and mostly met by coal-fired-plants, i.e. during weekends, nights and summers. Then, with adequate economic incentives, available gas-fired units may be switched. Consequently, the possibility of fuel-switching varies throughout the year depending on the season (winter or summer), the time of the week (day-of-week or week-end effects), and the period of the day (day or night).

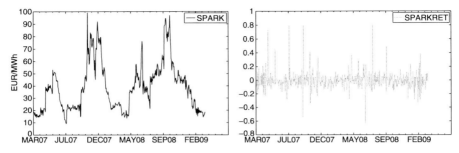

Fig. 2.10 Clean Spark Spread: Raw data (*left panel*) and logreturns (*right panel*) from March 09, 2007 to March 31, 2009

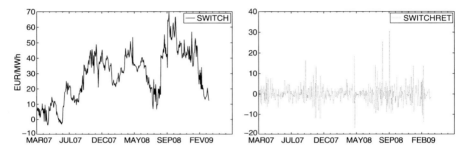

Fig. 2.11 Switch Price: Raw data (*left panel*) and logreturns (*right panel*) from March 09, 2007 to March 31, 2009

Clean Spark Spread   Second, the Clean Spark Spread (*clean spark*, expressed in Euro/MWh) represents the difference between the price of electricity at peak hours and the price of natural gas (*ngas*) used to generate that electricity, corrected for the energy output of the gas-fired plant and the costs of CO_2:

$$clean\ spark = elec - \left(ngas \times \frac{1}{\rho_{ngas}} + p_t \times EF_{ngas} \right) \quad (2.14)$$

with ρ_{ngas} the net thermal efficiency of a conventional gas-fired plant (i.e. 49% according to Reuters), and EF_{ngas} the CO_2 emissions factor of a conventional coal-fired power plant (i.e. 0.41 tCO_2/MWh according to Reuters). The Clean Spark Spread is shown in Fig. 2.10.

Switch Price   Third, the Switch Price (*switch*, expressed in Euro/MWh) represents the shadow price that allows equalizing the Clean Dark and Clean Spark spreads:

$$switch = \frac{cost_{ngas}/\text{MWh} - cost_{coal}/\text{MWh}}{tCO_{2coal}/\text{MWh} - tCO_{2ngas}/\text{MWh}} \quad (2.15)$$

with $cost_{ngas}$ the production cost of one MWh of electricity based on net CO_2 emissions of gas (expressed in EUR/MWh), $cost_{coal}$ the production cost of one MWh of electricity based on net CO_2 emissions of coal (expressed in Euro/MWh), tCO_{2coal}

Table 2.8 Cross-correlations between carbon (EUA) and electricity (DARK, SPARK, SWITCH) variables in logreturns transformation

	EUARET	DARKRET	SPARKRET	SWITCHRET
EUARET	1			
DARKRET	0.0687	1		
SPARKRET	−0.0455	0.1775	1	
SWITCHRET	0.1617	0.3075	−0.3586	1

the emissions factor (expressed in tCO_2/MWh) of a conventional coal-fired plant, and $tCO2_{ngas}$ the emissions factor (expressed in tCO_2/MWh) of a conventional gas-fired plant. The switch price is shown in Fig. 2.11. As long as the carbon price is below this switching price, coal plants are more profitable than gas plants—even after taking carbon costs into account.

Similarly to energy variables, we verify in Table 2.8 that the cross-correlations between the carbon price and electricity variables (taken in logreturns) do not exceed 0.6 in absolute value. Then, we can proceed to the regression analysis of carbon price drivers with electricity variables.

2.2.3.2 Regression Analysis

We run the following regression:

$$y_t = \alpha + \phi y_{t-1} + \beta_4 Clean\ Dark_t + \beta_5 Clean\ Spark_t + \beta_6 Switch_t + \varepsilon_t \quad (2.16)$$

with ϕ the parameter for the AR(1) process, and β_4 to β_6 the coefficient estimates for, respectively, *DARKRET*, *SPARKRET* and *SWITCHRET*. This regression can be estimated by using the R code in Table 2.9.

This code yields to the output reproduced in Table 2.10.

Besides the influence of the AR(1) process (significant at the 1% level), we uncover in this example the statistically significant impact of the Clean Spark Spread (*spark*) on carbon price changes at the 10% level. Neither the Clean Dark Spread, nor the Switch Price carry explanatory power in the model estimated according to Eq. (2.16). However, note that the sign of the β_4 coefficient obtained for the Clean Dark Spread (*dark*) is *negative*, which is consistent with the negative sign for the coal variable identified in Eq. (2.9). In addition, the user may experiment various lag specifications for electricity variables.

Lines 14–15 plot the ACF and PACF of the residuals from the electricity model in Eq. (2.16). By looking at Fig. 2.12, we conclude that the residuals of this regression are not autocorrelated. The p-value of the F-test statistic also confirms the joint significance of this regression.

Note that the factors which influence the fuel-switching behavior of power producers are further analyzed in Chap. 5.

Having detailed the channels through which energy prices are deemed to affect the price path of carbon, we consider in the next section a third category of carbon price drivers: extreme weather events.

Table 2.9   R Code: Linear Regression Model with Electricity Variables

```
1   path<-"C:/"
2   setwd(path)
3   library(dynlm)
4   data=read.csv("data_chapter2.csv",sep=",")
5   attach(data)
6   eua=data[,2]
7   dark=data[,8]
8   spark=data[,9]
9   switch=data[,10]
10  modelelectricity<-dynlm(formula=eua~L(eua)+dark+spark+switch)
11  summary(modelelectricity)
12  layout(matrix(1:4,2,2))
13  plot(modelelectricity)
14  acf(residuals(modelelectricity))
15  acf(residuals(modelelectricity),type='partial')
```

Table 2.10   R Output: Linear Regression Model with Electricity Variables

```
Call:
dynlm(formula = eua ~ L(eua) + dark + spark + switch)

Residuals:
      Min          1Q      Median          3Q         Max
-1.268e-17  -6.167e-19  -1.168e-19   4.896e-19   2.937e-17

Coefficients:
             Estimate Std. Error    t value Pr(>|t|)
(Intercept)  0.000e+00  9.685e-20  0.000e+00   1.0000
L(eua)       1.000e+00  2.679e-18  3.733e+17   <2e-16 ***
dark        -1.420e-18  1.089e-18 -1.304e+00   0.1930
spark        1.498e-18  8.746e-19  1.713e+00   0.0873 .
switch       4.100e-20  2.754e-20  1.489e+00   0.1371
---
Signif. codes:  0 '***' 0.001 '**' 0.01 '*' 0.05 '.' 0.1 ' ' 1

Residual standard error: 2.227e-18 on 524 degrees of freedom
Multiple R-squared:     1,      Adjusted R-squared:     1
F-statistic: 3.489e+34 on 4 and 524 DF,  p-value: < 2.2e-16
```

2.3   Extreme Weather Events

The influence of temperatures and weather events needs also to be accounted for when examining the carbon price drivers. We explain below the rationale behind these effects, along with an example.

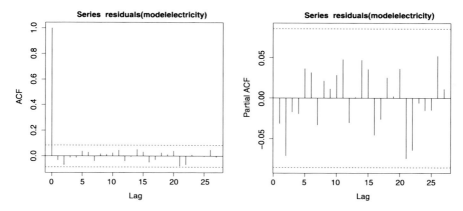

Fig. 2.12 Autocorrelation and partial-autocorrelation functions for the residuals from the model estimated in Eq. (2.16)

2.3.1 Relationship Between Temperatures and Carbon Prices

Taken together, cold and hot temperatures constitute the most important dimension of weather [45].[12] In turn, weather conditions are expected to impact the price path of carbon insofar as they influence energy demand. For instance, cold winters yield to increased use of heating, and thus extra need for power generation. Conversely, hot summers yield to extra need for air conditioning, which also increases electricity production. More power production will be achieved by more fuel use (associated with CO_2 emissions), and in fine there will be a need for more CO_2 allowances (with a positive effect on the CO_2 price). Therefore, we may consider a broad European temperatures index as a suitable carbon price driver.

To illustrate our point, we detail the functioning of the Metnext Weather index.[13] This national business-climate index is defined as the average daily temperature of the regions making up the country, weighted by the population of these regions. This methodology allows to derive a good approximation of the weight of regional economic activity. The index, expressed in Degree Celsius, is computed for eighteen countries[14] as follows:

$$\theta = \frac{\sum_{i=1}^{N} p_i \times \theta_i}{\sum_{i=1}^{N} p_i} \tag{2.17}$$

with N the number of regions in the country under consideration, p_i the population of region i, and θ_i the average temperature of region i during the month under consideration.

---

[12]Due to data limitations (and lack of availability at the European level), we do not consider here other potential weather events linked to rainfall or wind.

[13]See the website at the following address: http://www.weatherindices.com/index.

[14]Austria, Belgium, Germany, Denmark, Spain, Finland, France, United-Kingdom, Hungary, Ireland, Italy, the Netherlands, Norway, Poland, Portugal, Sweden, Slovakia and Slovenia.

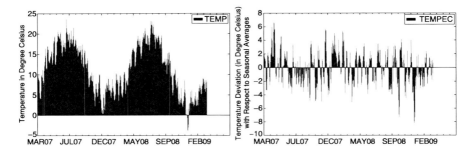

Fig. 2.13 Weather Events: European Temperatures Index (*left panel*) and Temperatures Deviation from Seasonal Average (*right panel*) in Degree Celsius from March 09, 2007 to March 31, 2009

Then, the CDC Climate Research[15] has extended this methodology to take into account the specificities of the carbon market. Their European temperatures index is equal to the average of the national temperature indices θ provided by Metnext Weather for the eighteen countries, weighted by the weight of each country in the total volume of distributed allowances:

$$T = \frac{\sum_{j=1}^{4} Q_j \times \theta_j}{\sum_{j=1}^{4} Q_j} \tag{2.18}$$

with T the monthly index, Q_j the number of allowances allocated by country j in its National Allocation Plan, and θ_j the national temperatures index of country j. An example of this European Temperatures Index (expressed in Degree Celsius) is displayed in the left panel of Fig. 2.13.

However, it is worth highlighting that the impact of temperatures on carbon prices is essentially nonlinear, i.e. it is possible to identify econometrically statistically significant effect of weather events only *above* or *below* a given threshold [2]. That is why, in the right panel of Fig. 2.13, we also reproduce temperatures deviation from decennial seasonal averages (expressed in Degree Celsius). The main insight behind this plot is that *unanticipated* temperatures events only can have an effect on carbon prices. If temperatures are conform to seasonal averages, then market participants (brokers, analysts, investors, ...) can properly anticipate them.

That is why we do not propose in the next section a regression analysis based on either the temperatures index, or the deviation from seasonal averages. These variables are not powerful enough (in statistical terms) to carry explanatory power concerning carbon price changes. Other econometric variables need to be constructed in order to capture the effects of extreme weather events.

---

[15]See the website at the following address: http://www.cdcclimat.com.

Table 2.11 R Code: Linear Regression Model with Extreme Weather Events

```
 1  path<-"C:/"
 2  setwd(path)
 3  library(stats)
 4  data=read.csv("data_chapter2.csv",sep=",")
 5  attach(data)
 6  eua=data[,2]
 7  tempcold=data[,11]
 8  temphot=data[,12]
 9  modelweather<-lm(eua~tempcold+temphot)
10  summary(modelweather)
11  layout(matrix(1:4,2,2))
12  plot(modelweather)
13  acf(residuals(modelweather))
14  acf(residuals(modelweather),type='partial')
```

2.3.2 Empirical Application

For the purpose of our empirical application, let us consider the following regression:

$$y_t = \alpha + \beta_7 Tempcold_t + \beta_8 Temphot_t + \varepsilon_t \qquad (2.19)$$

where $Tempcold_t$ is a dummy variable capturing the influence of cold temperatures (it is equal to one when the temperatures index in a given month is below decennial seasonal averages by -1.97 Degree Celsius), and $Temphot_t$ is a dummy variable capturing the influence of hot temperatures (it is equal to one when the temperatures index in a given month is above decennial seasonal averages by 1.47 Degree Celsius). As already described in Sect. 2.1.1, dummy variables appear as suitable tools for the investigation of weather effects on carbon prices. This regression may be estimated by using R in Table 2.11.

The results are shown in Table 2.12.

We may observe that the β_7 and β_8 coefficients, capturing the effects of $Tempcold_t$ and $Temphot_t$ respectively, are statistically significant at the 5% level. Therefore, we are able to confirm in this example that extreme temperatures events, i.e. temperatures events *above* or *below* decennial seasonal averages, have a statistically significant effect on carbon prices by following the economic logic explained in Sect. 2.3.1.

The p-value of the F-test statistic confirms that the joint significance of the results is accepted. As shown by the ACF and PACF functions in Fig. 2.14, we can also accept the null hypothesis that residuals of the regression are not autocorrelated.

Table 2.12 R Output: Linear Regression Model with Extreme Weather Events

```
Call:
lm(formula = eua ~ tempcold + temphot)

Residuals:
     Min       1Q    Median       3Q      Max
-0.10435 -0.01521 -0.00075  0.01435  0.54939

Coefficients:
              Estimate Std. Error t value Pr(>|t|)
(Intercept) -0.008816   0.004597  -1.918   0.0557 .
tempcold     0.009256   0.004621   2.003   0.0457 *
temphot      0.009561   0.004612   2.073   0.0386 *
---
Signif. codes:  0 '***' 0.001 '**' 0.01 '*' 0.05 '.' 0.1 ' ' 1

Residual standard error: 0.03611 on 526 degrees of freedom
Multiple R-squared: 0.009094,   Adjusted R-squared: 0.005327
F-statistic: 2.414 on 2 and 526 DF,  p-value: 0.09047
```

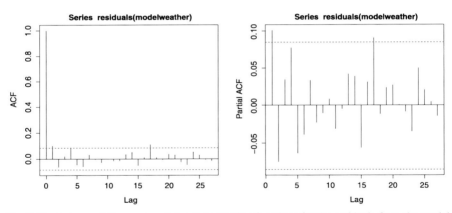

Fig. 2.14 Autocorrelation and partial-autocorrelation functions for the residuals from the model estimated in Eq. (2.19)

To conclude this chapter, we invite the reader to combine all the carbon price drivers considered here by using the linear regression model, in the spirit of the practical examples developed above, or by using GARCH models as introduced in the next chapter. As shown by previous academic literature [2], the influence of such drivers (mainly energy variables and extreme temperatures events) varies depending on the period under consideration and the influence of institutional decisions (which may also be accounted for by considering sub-samples).

Box 2.5 (Homoskedasticity Tests) This chapter has not dealt with the notion of non-constant variance (i.e. heteroskedasticity), which is gradually introduced in Chap. 3.

We briefly present below the main statistical tests to detect the presence of heteroskedasticity in the residuals of regression models.

According to the White test, a general test of homoskedasticity may be written as:

$$\hat{\varepsilon}_t^2 = a_0 + a_1 X_{t-1} + b_1 X_{t-1}^2 + \cdots + a_p X_{t-p} + b_p X_{t-p}^2 + v_t \qquad (2.20)$$

with $\hat{\varepsilon}_t$ the estimated residuals, and X_{t-p} the lagged endogenous variable. The test is based on the Lagrange multiplier (LM) with T the number of observations, and R^2 the R-squared of Eq. (2.20). If $TR^2 < \chi^2(2p)$, the null hypothesis of homoskedasticity is accepted (see Hamilton (1996, [33]) for more details).

According to the Engle ARCH test, the null hypothesis of homoskedasticity can also be tested by running the following regression:

$$\hat{\varepsilon}_t^2 = \alpha_0 + \sum_{i=1}^{l} \alpha_i \hat{\varepsilon}_{t-1}^2 \qquad (2.21)$$

If $TR^2 < \chi^2(l)$, the null hypothesis of homoskedasticity is accepted.

Homoskedasticity tests of the residuals can be performed in R with the following commands:

```
1 library(lmtest)
2 library(vars)
3 model<-lm(eua~ar(1)+ar(p))
4 bptest(model,~ar(1)*ar(p)+I(ar(1)^2)+I(ar(p)^2))
5 model<-VAR(eua,p=2)
6 arch(model)
```

Lines 1 and 2 load the relevant libraries. Lines 3 and 4 compute the White test as a special case of the `bptest()` command, by including all regressors and the squares/cross-products in the auxiliary regression. Lines 5 and 6 compute the Engle ARCH test in the framework of a VAR(p) (see Chap. 4).

Appendix: BEKK MGARCH Modeling with CO_2 and Energy Prices

Based on the Baba-Engle-Kraft-Kroner ([11, 30], BEKK) specification of the Multivariate GARCH (MGARCH) model, we highlight in this section the transmission of price volatility among carbon (EUA) and energy (oil, gas, coal) markets. Note that the univariate GARCH modelling framework is detailed in Chap. 3. The data for this section is available at: http://sites.google.com/site/jpchevallier/publications/books/springer/data_chapter2_appendix.csv.

The BEKK MGARCH Model   The class of multivariate GARCH models that we study stems from the contributions of Bollerslev, Engle and Wooldridge ([11], 1988), who provided with the VEC-GARCH model a straightforward extension of univariate GARCH models. We examine more particularly the following BEKK MGARCH model [30]:

$$H_t = CC' + \sum_{j=1}^{q}\sum_{k=1}^{K} A'_{kj} r_{t-j} r'_{t-j} A_{kj} + \sum_{j=1}^{p}\sum_{k=1}^{K} B'_{kj} H_{t-j} B_{kj} \tag{2.22}$$

where A_{kj}, B_{kj}, and C are $N \times N$ parameter matrices, and C is lower triangular to ensure the positive definiteness of H_t. Note the BEKK model is covariance stationary if and only if the eigenvalues of $\sum_{j=1}^{q}\sum_{k=1}^{K} A_{kj} \otimes A_{kj} + \sum_{j=1}^{p}\sum_{k=1}^{K} B_{kj} \otimes B_{kj}$ are less than one in modulus, with $\otimes$ the notation for Kronecker products. Due to the computational burden involved by the estimation of a full BEKK model,[16] we restrict the number of parameters by implementing the following "diagonal BEKK" MGARCH model:

$$H_t = CC' + A' r_{t-1} r'_{t-1} A + D E[A' r_{t-1} r'_{t-1} A \mid \Im_{t-2}] D \tag{2.23}$$

In Eq. (2.23), we now model the conditional variances and covariances of certain linear combinations of the vector of price returns r_t.

Estimation with R   The R code to estimate the BEKK MGARCH model is given in Table 2.13. Note the MGARCH BEKK package for R may be downloaded at: http://www.quantmod.com/download/mgarchBEKK/.

Table 2.13   R Code: Estimating the BEKK MGARCH model

```
1  path<-"C:/"
2  setwd(path)
3  library(mgarchBEKK)
4  data=read.csv("data_chapter2_appendix.csv",sep=",")
5  attach(data)
6  eua=data[,2]
7  oil=data[,3]
8  gas=data[,4]
9  coal=data[,5]
10 mgarch.mat<-data.frame(eua,oil,gas,coal)
11 model.BEKK<-mvBEKK.est(mgarch.mat,order=c(1,1,1,1),
   params=NULL)
12 mvBEKK.diag(model.BEKK)
```

---

[16] As the number of parameters $(p+q)KN^2 + N(N+1)/2$ increases, it may be difficult to obtain convergence during the estimation.

Table 2.14 R Output: Estimating the BEKK MGARCH model

```
          Number of estimated series :   4
          Length of estimated series :   529
          Estimation Time            :   7.7362
          Total Time                 :   7.760683
          BEKK order                 :   1 1 1 1
          Eigenvalues                :   1.371115 1.017605 0.9884797 0.72346 0.6930659 0.5845497 0.506614
                                         0.4460167 0.3941953 0.2683696 0.1986257 0.08505935 0.05234043
                                         0.03286894 0.007029388 0.004263158
          aic                        :   -5351.701
          unconditional cov. matrix  :   0.001148309 0.0001154143 -0.0006690489 0.0001950053 0.0001154143
                                         0.0006749642 -0.0004209732 -0.0001267162 -0.0006690489 -0.0004209732
                                         0.001150107 9.731776e-05 0.0001950053 -0.0001267162 9.731776e-05
                                         0.0001509592
          var(resid 1 )              :   0.9853412
          mean(resid 1 )             :   0.02929001
          var(resid 2 )              :   0.956688
          mean(resid 2 )             :   -0.001323343
          var(resid 3 )              :   1.045084
          mean(resid 3 )             :   0.01247064
          var(resid 4 )              :   1.034191
          mean(resid 4 )             :   0.05121331
          Estimated parameters       :

          C estimates:
              [,1]         [,2]         [,3]         [,4]
     [1,] -0.0181858 -0.002252538  0.017789029 -2.384224e-03
     [2,]  0.0000000  0.014754446 -0.022988227 -6.830455e-03
     [3,]  0.0000000  0.000000000  0.007573896  2.210108e-07
     [4,]  0.0000000  0.000000000  0.000000000 -1.180576e-04

          ARCH estimates:
              [,1]        [,2]        [,3]        [,4]
     [1,]  0.7660670  0.07154282  0.27820714  0.27219474
     [2,]  0.2332687  0.33612697 -0.03536674  0.02553457
     [3,] -0.2442194  0.01900365  0.01417112  0.22302539
     [4,]  0.1541698  0.05541739  0.05950522  0.18045794

          GARCH estimates:
              [,1]         [,2]         [,3]         [,4]
     [1,] -0.1066946471 -0.456940766 -0.56919659 -0.01460560
     [2,]  0.5081535812 -0.141011118  0.58304368 -0.27584249
     [3,] -0.0004665412  0.461722817  0.42903780 -0.01917917
     [4,]  0.0311903261 -0.003331396  0.02776824 -0.70484255

          asy.se.coef                :

          C estimates, standard errors:
              [,1]         [,2]         [,3]         [,4]
     [1,] 0.001856379 0.013672040 0.012979851 0.004585833
     [2,] 0.000000000 0.004089436 0.005244723 0.001519628
     [3,] 0.000000000 0.000000000 0.008788042 0.007997546
     [4,] 0.000000000 0.000000000 0.000000000 0.008027800

          ARCH estimates, standard errors:
              [,1]        [,2]        [,3]        [,4]
     [1,] 0.07108107 0.11079524 0.10873442 0.05337258
     [2,] 0.06340028 0.05735820 0.06200331 0.05674444
     [3,] 0.01243639 0.03171205 0.05404650 0.01870268
     [4,] 0.06916141 0.06702569 0.05706629 0.06795449

          GARCH estimates, standard errors:
              [,1]        [,2]        [,3]        [,4]
     [1,] 0.10369959 0.13957939 0.18168679 0.11921919
     [2,] 0.07110973 0.08714015 0.18288946 0.11336719
     [3,] 0.25998745 0.06305122 0.09347202 0.04677369
     [4,] 0.24299070 0.06075734 0.38037655 0.09695227
```

Fig. 2.15 Residual series for
the BEKK MGARCH model
estimated in Eq. (2.23)

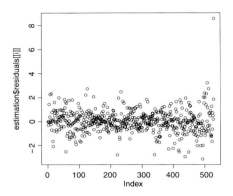

Lines 1 to 10 load the relevant library and data. Line 11 performs the estimation of the BEKK MGARCH model, which returns the output shown in Table 2.14.

The coefficients for the variance-covariance equations are generally significant for own- and cross-innovations, and significant for own- and cross-volatility spillovers in the individual price series of oil, gas, coal and CO$_2$, indicating the presence of strong ARCH and GARCH effects. In evidence, most of the estimated ARCH and GARCH coefficients are significant at the 1% level.

Line 12 provides the graph of the residuals, as pictured in Fig. 2.15. We cannot identify any pattern in the residual series, therefore we can conclude that the BEKK MGARCH model for carbon and energy markets is well-specified (without evidence of autocorrelation in the residuals).

The interested reader may refer to Chevallier (2012, [14]) for an extension of this analysis to constant conditional correlation ([10], CCC) and dynamic conditional correlation ([29, 31], DCC) MGARCH models. Also, the DCC MGARCH model has been applied to the relationship between EUAs and CERs [13].

Problems

Problem 2.1 (Unit Root Tests for Carbon Prices)

(a) Consider the time-series in Fig. 2.16: do you observe spikes (upside peaks followed by downside jumps)? are the trajectories formed by random oscillations? do the price series evolve around a random trend? does volatility seem higher and spikes more frequent during some specific periods?

(b) Consider the time-series in Fig. 2.17: do you observe correlation peaks following a specific pattern? which prices between day d and day $d - n$ seem more correlated than prices with other lags?

(c) Consider the Tables 2.15 and 2.16, which report the ADF, PP, and KPSS tests: what do you conclude? for which time-series can H_0 be rejected? for which time-series can you not reject H_0? detail the methodology to transform non-stationary variables to stationary, and the possible bias induced if the process is TS or DS.

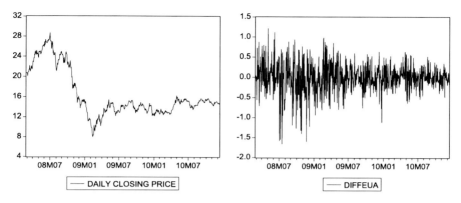

Fig. 2.16 Raw time-series (*left panel*) and logreturns (*right panel*) of the ECX EUA Futures Contract

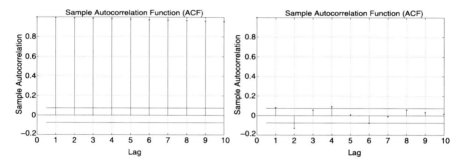

Fig. 2.17 ACF functions of the ECX EUA Futures contract in raw (*left panel*) and logreturn (*right panel*) forms

Problem 2.2 (Calibration of Mean-Reversion Models for Carbon Prices) The simplest mean-reversion model is the so-called AR(1) model:

$$x_{t+1} = \rho x_t + \beta + e_{t+1} \qquad (2.24)$$

where the (e_t) are i.i.d. with law $N(0, \sigma^2)$.

The calibration proceeds in two steps:

1. the coefficients ρ and β are determined by OLS regression of x_{t+1} on x_t;
2. σ is computed as the standard deviation of the residuals of the regression.
    (a) How would you calibrate this model to the CO_2 price series?
    (b) Consider the results presented in Table 2.17: what do you observe?
    (c) Consider the diagnostic tests presented in Table 2.18: what can you conclude concerning the hypothesis H_0 that the residuals are uncorrelated using the Box-Pierce test? concerning the hypothesis H_0 that the residuals follow a Gaussian law using the Jarque-Bera test? concerning the hypothesis H_0 that the residuals present heteroskedasticity using the Engle ARCH test?

Table 2.15　Unit root tests for the EUA_t variable

Null Hypothesis: EUA has a unit root

Exogenous: None

Lag Length: 0 (Automatic based on SIC, MAXLAG = 19)

	t-Statistic	Prob.*
Augmented Dickey-Fuller test statistic	−0.9301	0.3135
Test critical values:		
1% level	−2.5682	
5% level	−1.9412	
10% level	−1.6164	

*MacKinnon (1996, [40]) one-sided p-values

Null Hypothesis: EUA has a unit root

Exogenous: None

Bandwidth: 1 (Newey-West using Bartlett kernel)

	Adj. t-Stat	Prob.*
Phillips-Perron test statistic	−0.9206	0.3174
Test critical values:		
1% level	−2.5682	
5% level	−1.9412	
10% level	−1.6164	

*MacKinnon (1996, [40]) one-sided p-values

Null Hypothesis: EUA is stationary

Exogenous: Constant

Bandwidth: 22 (Newey-West using Bartlett kernel)

	LM-Stat.
Kwiatkowski-Phillips-Schmidt-Shin test statistic	1.5748
Asymptotic critical values*:	
1% level	0.7390
5% level	0.4630
10% level	0.3470

*Kwiatkowski-Phillips-Schmidt-Shin (1992, [39])

Note: ADF and PP: Model 1 (without intercept or trend) has been chosen. KPSS: Model 2 without trend has been chosen

Table 2.16 Unit root tests for the $D(EUA_t)$ variable

Null Hypothesis: D(EUA) has a unit root

Exogenous: Constant

Lag Length: 1 (Automatic based on SIC, MAXLAG = 19)

	t-Statistic	Prob.*
Augmented Dickey-Fuller test statistic	−20.4963	0.0000
Test critical values:		
1% level	−3.4393	
5% level	−2.8653	
10% level	−2.5688	

*MacKinnon (1996, [40]) one-sided p-values

Null Hypothesis: D(EUA) has a unit root

Exogenous: Constant

Bandwidth: 2 (Newey-West using Bartlett kernel)

	Adj. t-Stat	Prob.*
Phillips-Perron test statistic	−24.9767	0.0000
Test critical values:		
1% level	−3.4393	
5% level	−2.8653	
10% level	−2.5688	

*MacKinnon (1996, [40]) one-sided p-values

Null Hypothesis: D(EUA) is stationary

Exogenous: Constant

Bandwidth: 1 (Newey-West using Bartlett kernel)

	LM-Stat.
Kwiatkowski-Phillips-Schmidt-Shin test statistic	0.1408
Asymptotic critical values*:	
1% level	0.7390
5% level	0.4630
10% level	0.3470

*Kwiatkowski-Phillips-Schmidt-Shin (1992, [39])

Note: ADF and PP: Model 1 (without intercept or trend) has been chosen. KPSS: Model 2 without trend has been chosen

Table 2.17 Estimation of the mean-reversion model

Dependent Variable: DIFFEUA

Method: Least Squares

Sample (adjusted): 2/28/2008 12/09/2010

Included observations: 711 after adjustments

Variable	Coefficient	Std. Error	t-Statistic	Prob.
C	−0.0082	0.0143	−0.5742	0.5660
DIFFEUA(-1)	0.0620	0.0374	1.6564	0.0981
R-squared	0.0038	Mean dependent var		−0.0087
Adjusted R-squared	0.0024	S.D. dependent var		0.3840
S.E. of regression	0.3836	Akaike info criterion		0.9244
Sum squared resid	104.3328	Schwarz criterion		0.9372
Log likelihood	−326.6300	F-statistic		2.7438
Durbin-Watson stat	1.9850	Prob(F-statistic)		0.0980

Table 2.18 Diagnostic tests of the mean-reversion model

Diagnostic tests	
Box-Pierce test	0.8470
Jarque-Bera test	0.0001
Engle ARCH test	0.0034

Note: The Box-Pierce test is computed with 20 lags. The p-value of the Jarque-Bera test is given. The Engle ARCH test is computed with one lag

(d) How would you correct the potential mis-specification problems in the model estimated based on your answer to question (c)?

References

1. Alberola E, Chevallier J (2009) European carbon prices and banking restrictions: evidence from Phase I (2005–2007). Energy J 30:51–80
2. Alberola E, Chevallier J, Cheze B (2008) Price drivers and structural breaks in European carbon prices 2005-07. Energy Policy 36:787–797
3. Alberola E, Chevallier J, Cheze B (2008) The EU emissions trading scheme: the effects of industrial production and CO$_2$ emissions on European carbon prices. Int Econ 116:93–125
4. Alberola E, Chevallier J, Cheze B (2009) Emissions compliances and carbon prices under the EU ETS: a country specific analysis of industrial sectors. J Policy Model 31:446–462
5. Bai J, Perron P (1997) Estimation of a change point in multiple regression models. Rev Econ Stat 79:551–563
6. Bai J, Perron P (1998) Estimating and testing linear models with multiple structural changes. Econometrica 66:47–78
7. Bai J, Perron P (2003) Computation and analysis of multiple structural change models. J Appl Econom 18:1–22

8. Banerjee A, Dolado JJ, Galbraith JW, Hendry DF (1993) Cointegration, error correction, and the econometric analysis of non-stationary data. Oxford University Press, Oxford

9. Benz E, Trück S (2009) Modeling the price dynamics of CO_2 emission allowances. Energy Econ 31:4–15

10. Bollerslev T (1990) Modelling the coherence in the short-run nominal exchange rates: a multivariate generalized ARCH model. Rev Econ Stat 72:498–505

11. Bollerslev T, Engle RF, Wooldridge JM (1988) A capital asset pricing model with time-varying covariances. J Polit Econ 96:116–131

12. Bunn DW, Fezzi C (2009) Structural interactions of European carbon trading and energy prices. J Energy Markets 2:53–69

13. Chevallier J (2011) Anticipating correlations between EUAs and CERs: a dynamic conditional correlation GARCH model. Econ Bull 31:255–272

14. Chevallier J (2012) Macroeconomics, finance, commodities: Interactions with carbon markets in a data-rich model. Econ Model 28:557–567

15. Chevallier J, Ielpo F, Mercier L (2009) Risk aversion and institutional information disclosure on the European carbon market: a case-study of the 2006 compliance event. Energy Policy 37:15–28

16. Christiansen AC, Arvanitakis A, Tangen K, Hasselknippe H (2005) Price determinants in the EU emissions trading scheme. Climate Pol 5:15–30

17. Conrad C, Rittler D, Rotfuss W (2011) Modeling and explaining the dynamics of European Union Allowance prices at high-frequency. Energy Econ. doi:10.1016/j.eneco.2011.02.011

18. Convery FJ (2009) Reflections—the emerging literature on emissions trading in Europe. Rev Environ Econ Policy 3:121–137

19. Convery FJ, Redmond L (2007) Market and price developments in the European Union emissions trading scheme. Rev Environ Econ Policy 1:88–111

20. Daskalakis G, Markellos RN (2009) Are electricity risk premia affected by emission allowance prices? Evidence from the EEX, Nord Pool and Powernext. Energy Policy 37:2594–2604

21. Daskalakis G, Psychoyios D, Markellos RN (2009) Modelling CO_2 emission allowance prices and derivatives: evidence from the European trading scheme. J Bank Finance 33:1230–1241

22. Dalgaard P (2002) Introductory statistics with R. Springer statistics and computing. Springer, New York/Dordrecht/Heidelberg/London

23. Delarue E, D'haeseleer WD (2007) Price determination of ETS allowances through the switching level of coal and gas in the power sector. Int J Energy Res 31:1001–1013

24. Dickey DA, Fuller WA (1979) Distribution of the estimators for autoregressive time series with a unit root. J Am Stat Assoc 74:427–431

25. Dickey DA, Fuller WA (1981) Likelihood ratio statistics for autoregressive time series with a unit root. Econometrica 49:1057–1072

26. Ellerman AD, Buchner BK (2007) The European Union emissions trading scheme: origins, allocation, and early results. Rev Environ Econ Pol 166–187

27. Ellerman AD, Buchner BK (2008) Over-allocation or abatement? A preliminary analysis of the EU ETS based on the 2005–06 emissions data. Environ Resour Econ 41:267–287

28. Ellerman AD, Convery F, De Perthuis C (2010) Pricing carbon: the European Union emissions trading scheme. Cambridge University Press, Cambridge

29. Engle RF (2002) Dynamic conditional correlation: a simple class of multivariate generalized autoregressive conditional heteroskedasticity models. J Bus Econ Stat 20:39–350

30. Engle RF, Kroner KF (1995) Multivariate simultaneous generalized ARCH. Econom Theory 11:122–150

31. Engle RF, Sheppard K (2001) Theoretical and empirical properties of dynamic conditional correlation MGARCH. NBER working paper No 8554, National Bureau of Economic Research, Massachusetts, USA

32. Gujarati DN (2004) Basic econometrics, 4th edn. McGraw-Hill, New York

33. Hamilton JD (1996) Time series analysis, 2nd edn. Princeton University Press, Princeton

34. Harvey AC (1993) Time series models, 2nd edn. Harvester Wheatsheaf, New York

35. Hintermann B (2010) Allowance price drivers in the first phase of the EU ETS. J Environ Econ Manag 59:43–56

36. Kanen JLM (2006) Carbon trading and pricing. Environmental Finance Publications, London
37. Keppler JH, Mansanet-Bataller M (2010) Causalities between CO$_2$, electricity, and other energy variables during phase I and phase II of the EU ETS. Energy Policy 38:3329–3341
38. Kleiber C, Zeileis A (2008) Applied econometrics with R. Springer series use R! Springer, New York/Dordrecht/Heidelberg/London
39. Kwiatkowski DP, Phillips PCB, Schmidt P, Shin Y (1992) Testing the null hypothesis of stationarity against the alternative of the unit root: how sure are we that economic time series are non stationary? J Econom 54:159–178
40. MacKinnon JG (1996) Numerical distribution functions for unit root and cointegration tests. J Appl Econom 11:601–618
41. Mansanet-Bataller M, Pardo A (2009) Impacts of regulatory announcements on CO$_2$ prices. J Energy Mark 2:77–109
42. Mansanet-Bataller M, Pardo A, Valor E (2007) CO$_2$ prices, energy and weather. Energy J 28:73–92
43. Paolella MS, Taschini L (2008) An econometric analysis of emission allowance prices. J Bank Finance 32:2022–2032
44. Phillips PCB, Peron P (1988) Testing for a unit root in time series regression. Biometrika 75:335–346
45. Roll R (1984) Orange juice and weather. Am Econ Rev 74:861–880
46. Shumway RH, Stoffer DS (2011) Time series analysis and its applications with R examples, 3rd edn. Springer texts in statistics. Springer, New York/Dordrecht/Heidelberg/London
47. Zeileis A, Kleiber C (2005) Validating multiple structural change models—a case study. J Appl Econom 20:685–690

Chapter 3
Link with the Macroeconomy

Abstract This chapter is dedicated to the connection between carbon prices and macroeconomic risk factors, besides the other determinants linked to energy and institutional variables studied in previous chapters. Several variables from the stock and bond markets are first studied, along with their influence on the carbon market. Second, macroeconomic, financial and commodity indicators are introduced by the means of factor models. Third, the relationship between carbon prices and industrial production is investigated based on nonlinearity tests, self-exciting threshold autoregressive models, smooth transition autoregressive models and Markov regime-switching models. Overall, the results show that carbon allowances form a very specific market among energy commodities, and that the interactions with the macroeconomy differ depending on several parameters.

3.1 Stock and Bonds Markets

This section studies the empirical relationship between carbon prices and the main macroeconomic risk factors from the stock and bond markets: Treasury-Bill yields, equity dividend yields, the excess of 'junk' bond yields over high-quality corporate bond yields, and the excess return on a broadly diversified portfolio of commodities. Before investigating this relationship, we detail the GARCH modeling framework in econometrics applied to the carbon price.

3.1.1 GARCH Modeling of the Carbon Price

Following [9] and [59], carbon prices may be adequately modeled by resorting to the class of GARCH(p, q) models. Hence, we specify the most parsimonious specification of the AR(1)-GARCH(1, 1) model:

$$r_t = \mu + \rho_1 r_{t-1} + \varepsilon_t, \quad \varepsilon_t \mid \Omega_{t-1} \sim \text{i.i.d.}(0, \sigma_t^2) \tag{3.1}$$

$$\sigma_t^2 = \omega + \alpha \varepsilon_{t-1}^2 + \beta \sigma_{t-1}^2 \tag{3.2}$$

where r_t is the EUA futures contract return for day t, μ is a constant, and ε_t is the error term process with mean zero and conditional variance σ_t^2. The AR(1)-GARCH(1, 1) model is estimated by Quasi Maximum Likelihood (QML) by using Bollerslev-Wooldridge robust standard errors. Innovations follow a normal distribution. The BHHH optimization algorithm (Berndt et al. (1974, [13])) allows to obtain numerically the parameter estimates.

Box 3.1 (ARCH(q) and GARCH(p, q) Models)   When the data generating process exhibits nonlinearity in variance (i.e. when the variance depends on the entire information set including time variation), it is possible to rely on econometric models based on an endogenous parameterization of the conditional variance.
Let ε_t be the error term:

$$E(\varepsilon_t|\varepsilon_{t-1}) = 0 \tag{3.3}$$
$$V(\varepsilon_t|\varepsilon_{t-1}) = \sigma_t^2 \tag{3.4}$$

with $\varepsilon_{t-1} = (\varepsilon_{t-1}, \varepsilon_{t-2}, \ldots)$, and σ_t^2 the conditional variance of ε_t which can vary through time (contrary to ARMA models). Note we can also refer to a standardized process z_t instead of ε_t:

$$z_t = \varepsilon_t(\sigma_t^2)^{-1/2} = \frac{\varepsilon_t}{\sigma_t} \tag{3.5}$$

which has a conditional expectation equal to zero and an unitary variance.
Following the seminal contribution by Engle (1982, [35]), the AutoRegressive Conditional Heteroskedasticity (ARCH) model relies on a quadratic parameterization of the conditional variance. σ_t^2 is formulated as a linear function of the q past values of the squared innovations. The ARCH(q) process is given by:

$$\sigma_t^2 = \alpha_0 + \sum_{i=1}^{q} \alpha_i \varepsilon_{t-i}^2 = \alpha_0 + \alpha(L)\varepsilon_t^2 \tag{3.6}$$

with L the lag operator, $\alpha_0 > 0$, and $\alpha_i \geq 0 \; \forall i$. The constraints on the coefficients guarantee the positivity of the conditional variance. The variance is finite if $\sum_{i=1}^{q} \alpha_i < 1$.
This model allows to take into account volatility clustering (i.e. strong price variations followed by other strong price variations whose sign is not predictable).
Next, following Bollerslev ([14], 1986), the Generalized AutoRegressive Conditional Heteroskedasticity (GARCH) model introduces lagged values of the variance, which yields to a positive and parsimonious description of the lag structure. The GARCH(p, q) process is defined as:

$$\sigma_t^2 = \alpha_0 + \sum_{i=1}^{q} \alpha_i \varepsilon_{t-i}^2 + \sum_{j=1}^{p} \beta_j \sigma_{t-j}^2 = \alpha_0 + \alpha(L)\varepsilon_t^2 + \beta(L)\sigma_t^2 \tag{3.7}$$

with $\alpha_0 > 0$, $\alpha_i \geq 0 \; \forall i$, $\beta_j \geq 0 \; \forall j$. Similarly, the constraints on the coefficients guarantee the positivity of the conditional variance. Note that when $p = 0$, the GARCH(p, q) model reduces to the ARCH(q) model.

Table 3.1   R Code: GARCH Modeling

```
1   path<-"C:/"
2   setwd(path)
3   library(fGarch)
4   data=read.csv("data_chapter3.csv",sep=",")
5   attach(data)
6   eua=data[,2]
7   modelgarch<-garchFit( ~ garch(1,1), data = eua)
8   summary(modelgarch)
9   plot(modelgarch)
```

In GARCH models, since the conditional variance has a quadratic form, volatility can be more important following a decrease than following an increase of the process. This symmetric property of the conditional variance may not be fully satisfactory. Besides, due to the non-negativity constraints on the parameters, a shock has the same positive effect on volatility, which is incompatible with random or cyclical volatility processes. Exponential Generalized AutoRegressive Conditional Heteroskedasticity (EGARCH) by Nelson (1991, [58]) and Threshold Generalized AutoRegressive Conditional Heteroskedasticity (TGARCH) by Zakoian (1994, [80]) models have been designed to cope with these restrictions (see Engle and Patton (2001, [39]), Engle (2001, [36]) and Engle (2003, [37]) for useful reviews on this topic).

This model can be estimated in R with the code provided in Table 3.1. The dataset used in this example can be downloaded at: http://sites.google.com/site/jpchevallier/publications/books/springer/data_chapter3.csv.

Lines 1 to 6 load the relevant library[1] and data. Line 7 performs the estimation of the GARCH(1, 1) model, as described in Eq. (3.1). The first R output is the GARCH numerical optimization procedure (not shown here to conserve space).

As shown in Table 3.2, Line 8 gives the GARCH coefficient estimates.

This Table contains many useful information concerning the goodness-of-fit of the GARCH model. Indeed, we can observe that the ARCH and GARCH coefficients are highly significant (at the 1% level). Besides, the residuals do not appear to be autocorrelated, as judged by the high p-values of the Ljung-Box test with different lags. Finally, the ARCH test results shows that the heteroskedasticity has been properly taken into account in the modeling of the EUA time series.

Line 9 provides more diagnostic plots, which can be found in Fig. 3.1. We confirm visually the absence of autocorrelation in the residuals (or in the standardized residuals). Besides, we can observe finally the aspect of the conditional variance of

---

[1]Note the R package 'fGarch' requires to install Rmetrics. See https://www.rmetrics.org/.

Table 3.2 R Output: GARCH coefficient estimates

```
Title:
 GARCH Modelling

Call:
 garchFit(formula = ~garch($1, 1$), data = eua)

Mean and Variance Equation:
 data ~ garch($1, 1$)
<environment: 0214a74c>
 [data = eua]

Conditional Distribution:
 norm

Coefficient(s):
        mu        omega      alpha1       beta1
4.9252e-04  1.3111e-09  8.4561e-02  9.4023e-01

Std. Errors:
 based on Hessian

Error Analysis:
        Estimate  Std. Error  t value Pr(>|t|)
mu     4.925e-04   1.001e-03    0.492    0.623
omega  1.311e-09   3.722e-06    0.000    1.000
alpha1 8.456e-02   1.854e-02    4.562 5.07e-06 ***
beta1  9.402e-01   1.439e-02   65.318  < 2e-16 ***
---
Signif. codes:  0 '***' 0.001 '**' 0.01 '*' 0.05 '.' 0.1 ' ' 1

Log Likelihood:
 1149.913    normalized:  2.173748

Description:
 Fri May 20 20:06:29 2011 by user: jchevall


Standardised Residuals Tests:
                             Statistic p-Value
 Jarque-Bera Test   R    Chi^2 9601.376  0
 Shapiro-Wilk Test  R    W     0.8878204 0
 Ljung-Box Test     R    Q(10) 10.37186  0.4084968
 Ljung-Box Test     R    Q(15) 15.50072  0.4159866
 Ljung-Box Test     R    Q(20) 21.48272  0.3692258
 Ljung-Box Test     R^2  Q(10) 0.1748325 1
 Ljung-Box Test     R^2  Q(15) 0.3461913 1
 Ljung-Box Test     R^2  Q(20) 0.457268  1
 LM Arch Test       R    TR^2  2.13203   0.999175

Information Criterion Statistics:
     AIC       BIC       SIC      HQIC
-4.332374 -4.300079 -4.332487 -4.319732
```

Fig. 3.1 GARCH diagnostic plots (from *left* to *right* and *top* to *bottom*): EUA Time Series, Conditional Standard Deviation, ACF of Observations, ACF of Squared Observations, Cross Correlation, Residuals, Standardized Residuals, ACF of Standardized Residuals, ACF of Squared Standardized Residuals, QQ Plot of Standardized Residuals

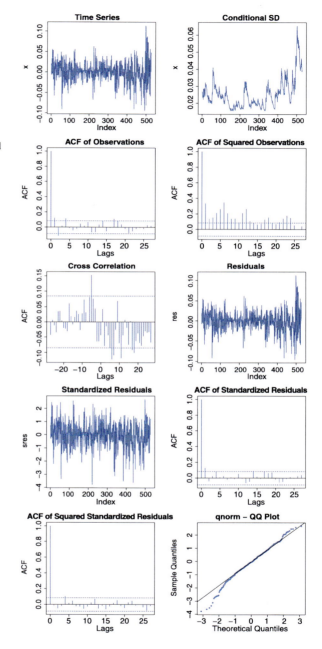

EUAs, as modeled by the GARCH(1, 1) process. Overall, this section has shown how the GARCH estimation procedure can be a precious tool to model adequately the volatility properties of carbon prices.

For more details on GARCH models, the interested reader may refer to Boller-slev, Chou and Kroner ([15], 1992), Bera and Higgins ([10], 1993), Bollerslev, Engle and Nelson ([16], 1994), and Li, Ling and McAleer ([53], 2002).

3.1.2 Relationship with Stock and Bond Markets

This section deals with the transmission of macroeconomic shocks to the carbon market.[2] This effect has been documented by Chevallier (2009, [26]), building on previous work for agricultural and energy futures markets by Bessembinder and Chan (1992, [12]), Bailey and Chan (1993, [4]), Fleming and Ostdiek (1999, [44]), and Sadorsky [63, 64]. The rationale behind the relationship between the carbon market and the macroeconomy stems from the necessary arbitrages made by industrials to adapt their productive capacities. When the demand for goods (and therefore real activity) is high, industrials need to produce and emit more CO_2 emissions. This extra demand for CO_2 allowances has a positive effect on the price of carbon, and conversely.[3]

The financial economics literature broadly considers four categories of macroeconomic risk factors:

1. dividend yields,
2. 'junk' bond yields,
3. Treasury bill yields,
4. market portfolio excess returns.

These indicators generally exhibit significant forecast power in futures markets (Fama and French [41, 43]). The existence of inter-relationships between stocks, bonds and commodities would thus suggest that there are common influences across these markets which lead towards the equilibrium price. Of course, these effects vary depending on the exposure of a given market to macroeconomic shocks.[4] We wish to assess whether these macroeconomic variables statistically influence the price of carbon. To do so, we detail each factor in relation to the carbon market.

3.1.2.1 Dividend Yields

As stated by Chevallier (2009, [26]), if futures commodity prices reflect the systematic risk embedded within the evolution of stock market conditions, we expect to identify a statistically significant relationship between carbon futures and the stock market.

---

[2]The spot carbon price series is not studied here due to the banking restrictions implemented between Phases I and II of the EU ETS (see Chap. 2 on this topic).

[3]This basic relationship will be explained in more details in the next sections of this chapter.

[4]For example, wood and heating oil are typically more sensitive to the business cycle than gold and silver, which exhibit little variation with the stock markets (and may be considered as safe havens in periods of market turbulence).

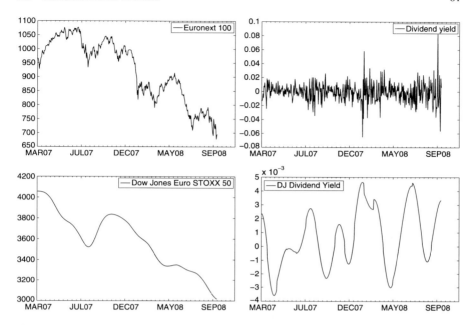

Fig. 3.2 Macroeconomic risk factors for stock markets (from *left* to *right* and *top* to *bottom*): Euronext 100, dividend yields on the Euronext 100, Dow Jones Euro Stoxx 50, dividend yields on the Dow Jones Euro Stoxx 50

As shown in Fig. 3.2, stock market variables can be approximated through the Euronext 100 or the Dow Jones Eurostoxx 50 indices. They represent the value-weighted portfolio of the most liquid stocks on the European exchanges. Their respective dividend yields, as shown in the right panels of Fig. 3.2, can be used for the econometric analysis. Dividend yields tend to be higher during periods of slow economic growth (Fama and French (1987, [40])).

3.1.2.2 Junk Bond Yields

Corporate bonds may be seen as less risky investments than stocks. The difference between low- and high-quality bonds represents typically the monetary compensation for risk required by investors to hold risky assets. In this context, 'junk' bonds variables can be defined as the excess of the yield between long term corporate bonds of different ratings quality. They are used as a proxy of default risk, and may be viewed as strong predictors of stock market returns (Fama and French (1989, [41])).

For instance, the top panel of Fig. 3.3 shows Moody's AAA- and BAA-rated corporate bonds in the left and middle panels, respectively. The junk bond yield is computed in the right panel by substracting the yields on BAA- and AAA-rated bonds (i.e. it represents the yield difference between the bonds of the lowest quality compared to that of the highest quality). This variable is expected to react quickly

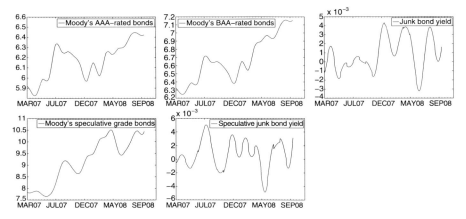

Fig. 3.3 Macroeconomic risk factors for bond markets (from *left* to *right* and *top* to *bottom*): Moody's corporate bonds rated AAA, Moody's corporate bonds rated BAA, junk bond yield for Moody's BAA-AAA bonds, Moody's speculative grade bonds, junk bond yield for Moody's speculative grade bonds

to market conditions, and we can test statistically whether there are feedback effects with the carbon market.

3.1.2.3 Treasury Bill Yields

Interest rates can be seen as a primary macroeconomic indicator. Their variation constitutes one of the most important indicator of the variation of systematic risk. It allows to track changes in the underlying business cycle, as macroeconomic policy channels extensively through the use of the increases/decreases in the basis points fixation of the interest rate.

As shown in the top panel of Fig. 3.4, the 90-day US Treasury-Bill (T-Bill) yield curve rate may be used as a proxy of current economic conditions. This variable, which is based on the closing market bid yields on actively traded US Treasury securities in the over-the-counter market, is used by market participants as the short-term discount rate linked to macroeconomic trends. Therefore, it tends to be high (low) during periods of economic growth (recessions). Alternatively, in the bottom panel of Fig. 3.4, the 2-year Euro area Government Benchmark bond yield from the European Central Bank can be used. The latter variables has the advantage of capturing changes in the macroeconomic environment specific to the EU ETS. Hence, we can also expect a positive relationship between returns on interest rates and carbon prices.

3.1.2.4 Market Portfolio Excess Returns

The excess return is typically computed as the difference between the return on a continuously compounded index of commodities and the continuously compounded

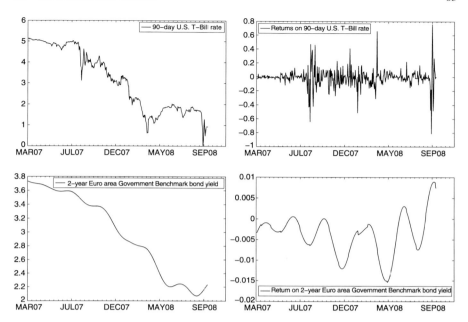

Fig. 3.4 Macroeconomic risk factors for interest rates (from *left* to *right* and *top* to *bottom*): 90-day US Treasury yield curve rate, return on the 90-day US Treasury yield curve rate, 2-year Euro Area Government Benchmark bonds yield, return on the 2-year Euro Area Government Benchmark bonds yield

90-day T-Bill rate. The choice of a commodity index as market portfolio is coherent with the view that carbon allowances form a new asset in the sphere of commodities (with the notable exception that the storage costs are virtually inexistent, as detailed in Chap. 5).

The top panel of Fig. 3.5 plots the Reuters/Jefferies Commodity Research Bureau (CRB) Futures index, which represents broad trends in overall commodity prices. It reflects changes in economic conditions that affect a basket of the most actively traded commodities in the metals, agricultural and energy sectors. Some commodities such as gold offer a safe haven during periods of market turbulence, hence the sign of the relationship with the carbon market depends on the cyclical or counter-cyclical nature of the index. In the bottom panel of Fig. 3.5, the Goldman Sachs commodity indicator (GSCI) follows the same logic as a broadly diversified portfolio of commodities, which allows to capture the influence of risk factors linked to global commodity markets.

3.1.2.5 Estimation Results from GARCH Models

The econometric strategy to identify the influence of macroeconomic risk determinants on the carbon market may be written as:

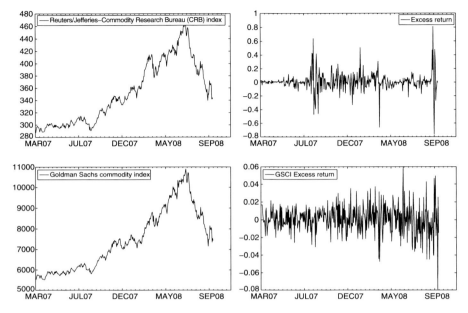

Fig. 3.5 Macroeconomic risk factors for commodity markets (from *left* to *right* and *top* to *bottom*): Reuters-CRB Futures index, excess return on the Reuters-CRB Futures index, Goldman Sachs commodity indicator, excess return on the Goldman Sachs commodity indicator

$$Y_t = \theta \mathbf{X_t'} + \varepsilon_t \tag{3.8}$$

$$\sigma_t^2 = \omega + \sum_{i=1}^{p} \alpha_i \varepsilon_{t-i}^2 + \sum_{j=1}^{q} \beta_j \sigma_{t-j}^2 \tag{3.9}$$

with σ_t^2 the conditional variance, which is function of a constant term ω, the ARCH term ε_{t-i}^2, and the GARCH term σ_{t-j}^2. The macroeconomic risk factors are included as exogenous regressors $\mathbf{X_t'}$ in the mean equation of the GARCH model.

To sum up the expected relationships between carbon futures and macroeconomic risk factors, we can expect that the stock and excess return variables are negatively related to the carbon market, while the T-Bill rate and 'junk' bond yields should be positively correlated with carbon prices (Chevallier (2009, [26])).

Box 3.2 (Further GARCH(p, q) Models) Various GARCH specifications can be tested in order to track the relationship between macroeconomic risk factors and the carbon market. Some of them seem particularly suited to detect the presence of mean reversion or leverage effects (Benz and Trueck (2009, [9]), Chevallier (2009, [26])).

The Component GARCH Model The standard GARCH(p, q) model may be rewritten as:

$$\sigma_t^2 = \overline{\omega} + \alpha(\varepsilon_{t-i}^2 - \overline{\omega}) + \beta(\sigma_{t-j}^2 - \overline{\omega}) \tag{3.10}$$

If the conditional variance shows mean reversion to $\overline{\omega}$, we may introduce the Component GARCH (CGARCH) model as follows:

$$\sigma_t^2 - m_t = \overline{\omega} + \alpha(\varepsilon_{t-i}^2 - \overline{\omega}) + \beta(\sigma_{t-j}^2 - \overline{\omega}) \tag{3.11}$$

$$m_t = \omega + \rho(m_{t-1} - \omega) + \phi(\varepsilon_{t-i}^2 - \sigma_{t-j}^2) \tag{3.12}$$

where ω is the time-varying long-run volatility, m_t the long-run component, and Eq. (3.11) describes the transitory component.

The Exponential GARCH Model Let us recall the Exponential GARCH (EGARCH) specification for the conditional variance proposed by Nelson (1991, [58]):

$$\log(\sigma_t^2) = \omega + \sum_{i=1}^{p} \alpha_i \left| \frac{\varepsilon_{t-i}}{\sigma_{t-i}} \right| + \sum_{j=1}^{q} \beta_j \log(\sigma_{t-j}^2) + \sum_{k=1}^{r} \gamma_k \frac{\varepsilon_{t-k}}{\sigma_{t-k}} \tag{3.13}$$

where γ tests for the presence of the leverage effect. Recall that the leverage effect implies a higher level of volatility associated to decreasing prices in the financial economics literature.

The GARCH-in-Mean Model If we introduce the log of the conditional variance in the mean equation, we get the GARCH(p, q)-in-Mean model (Engle et al. (1987, [38])):

$$Y_t = \theta \mathbf{X}_t' + \lambda \log(\sigma_t^2) + \varepsilon_t \tag{3.14}$$

The Power ARCH Model The Power ARCH (PARCH) specification allows to model the standard deviation rather than the variance, and may be written as follows (Ding et al. (1993, [32])):

$$\sigma_t^\delta = \omega + \sum_{i=1}^{p} \alpha_i (|\varepsilon_{t-i}| - \gamma_i \varepsilon_{t-i})^\delta + \sum_{j=1}^{q} \beta_j \sigma_{t-j}^\delta \tag{3.15}$$

with $\delta > 0$, $|\gamma_i| \leq 1 \ \forall i = 1, \ldots, \tau$, $\gamma_i = 0 \ \forall i > \tau$, $\tau < p$. δ is the power parameter of the standard deviation, and γ parameters to capture asymmetry up to order τ.

The Threshold GARCH Model Finally, the asymmetric TGARCH model by Zakoian (1994, [80]) may be written as:

$$\sigma_t^2 = \omega + \sum_{i=1}^{p} \alpha_i \varepsilon_{t-i}^2 + \sum_{j=1}^{q} \beta_j \sigma_{t-j}^2 + \sum_{k=1}^{r} \gamma_k \varepsilon_{t-k}^2 \Gamma_{t-k} \tag{3.16}$$

where $\Gamma_t = 1$ if $\varepsilon_t < 0$, and 0 otherwise. $\varepsilon_{t-i} > 0$ and $\varepsilon_{t-i} < 0$ denote, respectively, good and bad news.

By estimating various ARMA-GARCH models of carbon futures returns with macroeconomic variables,[5] Chevallier (2009, [26]) establishes that a weak link exists between the stock and bond markets on the one hand, and the carbon market on the other hand. Besides, the results do not show significant evidence in favor of the interest rates and global commodity markets. Interestingly, these results challenge the view by market analysts that macroeconomic risk factors have an immediate impact on the carbon market. Instead, Chevallier (2009, [26]) shows that only the default yield spread in the bond market and the dividend yields in the stock market have a statistically significant influence on carbon futures prices.

Therefore, previous econometric analyses exhibit a limited influence of macroeconomic risk factors on carbon prices, which channels through the stock market variable. The fact that carbon allowances constitute an easily storable commodity[6] may contribute to explain why we identify such a limited impact of stock and bond markets, and virtually no association with the T-Bill rate. In addition, the carbon price interacts primarily with other energy markets, as detailed in Chap. 2: price fundamentals are essentially a function of allowance supply fixed by the European Commission, and power demand arising from electric operators. This interplay with power producers' fuel-switching behavior may explain the direct link between carbon price changes and other energy price changes, and why the carbon market appears only remotely connected to macroeconomic variables. Last but not least, due to its specific institutional characteristic as an environmental market, we may also explain why that the carbon market is less sensitive to macroeconomic risk factors than the equity, bond and commodity markets.

Next, we investigate in further details the inter-relationships between the carbon market and various macroeconomic, financial and commodity indicators by using factor models.

3.2 Macroeconomic, Financial and Commodity Indicators

Factor models have become increasingly popular in applied econometrics since the contribution by Stock and Watson (2002, [65, 66]). This approach uses principal component analysis (PCA) in order to extract and summarize the information from large datasets, which can then be applied to the analysis of a given time series. The main advantage lies in the fact that the econometrician does not need to know *ex ante* which variables are supposed to have an effect on the time series under consideration (based on previous theoretical or empirical studies for instance). Instead, the econometrician takes an agnostic stance by building large datasets (in our setting, composed of macroeconomic, financial and commodity variables), by extracting this information into a limited number of factors, and by looking at their explanatory power for the returns of the phenomenon under study. These factor

---

[5]Note that the dataset for this study is not provided in this chapter.

[6]Compared to gold and silver for instance, industrials do not incur any storage cost for carbon allowances (see Chap. 5 for more details).

models (Fama and French (1992, [42])) can also be used in a vector auto-regressive framework (VAR),[7] as shown in our empirical application to EUAs.

3.2.1 Extracting Factors Based on Principal Component Analysis

Given a k-dimensional random variable $\mathbf{r} = (r_1, \ldots, r_k)'$ with covariance matrix $\sum_r$, principal component analysis consists in using a few linear combinations of r_i in order to explain the structure of $\sum_r$ (Tsay (2010, [78])).[8]

Let $\mathbf{w}_i = (w_{i1}, \ldots, w_{ik})'$ be a k-dimensional real-valued vector, with $i = 1, \ldots, k$. The linear combination of the random vector $\mathbf{r}$ writes:

$$y_i = \mathbf{w}_i' \mathbf{r} = \sum_{j=1}^{k} w_{ij} r_j \qquad (3.17)$$

where $\mathbf{r}$ are the returns of the k variables, and y_i is the return of the portfolio which assigns the weight w_{ij} to the jth variable. As stated above, the main insight behind PCA consists in finding the linear combinations $\mathbf{w}_i$, such that:

- y_i and y_j are uncorrelated for $i \neq j$;
- the variances of y_i are as large as possible.

According to the linear combination properties of random variables, we know that (Tsay (2010, [78])):

$$\mathrm{Var}(y_i) = \mathbf{w}_i' \sum_r \mathbf{w}_i, \quad i = 1, \ldots, k \qquad (3.18)$$

$$\mathrm{Cov}(y_i, y_j) = \mathbf{w}_i' \sum_r \mathbf{w}_j, \quad i, j = 1, \ldots, k \qquad (3.19)$$

Therefore, the first principal component of $\mathbf{r}$ is the linear combination $y_1 = \mathbf{w}_1' \mathbf{r}$ which maximizes $\mathrm{Var}(y_1)$ subject to $\mathbf{w}_1' \mathbf{w}_1 = 1$. The second principal component of $\mathbf{r}$ is the linear combination $y_2 = \mathbf{w}_2' \mathbf{r}$ which maximizes $\mathrm{Var}(y_2)$ subject to $\mathbf{w}_2' \mathbf{w}_2 = 1$ and $\mathrm{Cov}(y_2, y_1) = 0$. The ith principal component of $\mathbf{r}$ is the linear combination $y_i = \mathbf{w}_i' \mathbf{r}$ which maximizes $\mathrm{Var}(y_i)$ subject to $\mathbf{w}_i' \mathbf{w}_i = 1$ and $\mathrm{Cov}(y_i, y_j) = 0$ for $j = 1, \ldots, i - 1$.

The interested reader can refer to Tsay (2010, [78]) for the resolution of this problem, based on the spectral decomposition of the covariance matrix $\sum_r$. The result is that the proportion of the total variance in $\mathbf{r}$ explained by the ith principal component is the ratio between the ith eigenvalue and the sum of all eigenvalues of $\sum_r$.

In empirical applications, it is also useful to compute the cumulative proportion of total variance explained by the first i principal components. By following this

---

[7] See Chap. 4 for a description of the basic VAR model.

[8] Note that PCA can also be used for the correlation matrix ρ_r of $\mathbf{r}$.

Fig. 3.6 EUA Spot and
Futures Prices from April
2008 to June 2009

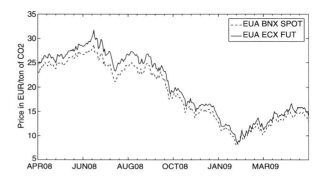

procedure, a small number of factors (but with a large corresponding cumulative proportion) can be selected, as shown in the next section for EUAs.

3.2.2 Factor-Augmented VAR Analysis Applied to EUAs

We present first the data used, second the econometric framework, and third the estimation results.

3.2.2.1 Data

First, we consider the time series of EUA spot and futures prices from April 2008 to June 2009, as shown in Fig. 3.6.[9]

Second, we extract two factors from a large database composed of macroeconomic, financial and commodity indicators by following the PCA approach detailed above. The dataset includes a large number of time series related to stocks, bonds, industrial production, market and price indices, exchange rates and various monetary aggregates. It also contains commodity and energy related variables which are deemed to interact with carbon markets.[10]

The factor loadings, which represent the contribution of each time series to the factors, are shown in Fig. 3.7. They show, by and large, that macroeconomic and financial time series provide the largest contribution to the first factor, while commodities account for the largest weight in the second factor (see Chevallier (2011, [27] for the list of variables). Together, the two factors account for approximately 50% of the total variance contained in the dataset.

Next, we specify a Factor-Augmented VAR model (FAVAR) in the spirit of Bernanke et al. (2005, [11]) to instrument the relationship between the returns on carbon prices and the factors.

---

[9]The data for this chapter is not available for download.

[10]See Chevallier (2011, [27]) for the description of the time series which have been gathered in a large database to represent the broad economic environment.

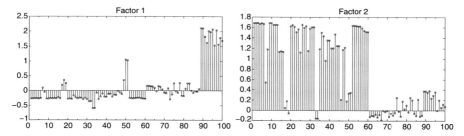

Fig. 3.7 Factor loadings extracted from the large dataset of macroeconomic and financial (*left panel*) and commodity (*right panel*) variables with the PCA

3.2.2.2 The Factor-Augmented VAR Model

Noting y_t an $M \times 1$ vector of time-series variables, and denoting y_{it} a particular variable, the FAVAR builds on the dynamic factor model structure commonly formulated in state-space form (Stock and Watson (2002, [66])):

$$y_{it} = \lambda_{0i} + \lambda_i f_t + \gamma_i r_t + \varepsilon_{it} \tag{3.20}$$

$$f_t = \Phi_1 f_{t-1} + \cdots + \Phi_p f_{t-p} + \varepsilon_t^f \tag{3.21}$$

for $i = 1, \ldots, M$, where f_t is defined as a $q \times 1$ vector of unobserved latent factors (where $q \ll M$) which contains information extracted from all the M variables assumed to drive the dynamics of the economy, λ_i is an $1 \times q$ matrix of so-called factor loadings, λ_{0i} represents the intercept for every dependent variable, r_t is a $k_r \times 1$ vector of observed variables, ε_{it} is i.i.d. $N(0, \sigma_i^2)$ and ε_t^f is i.i.d. $N(0, \sum^f)$.

The FAVAR extends the state equation for the factors to allow for r_t to have a VAR form:

$$\begin{pmatrix} f_t \\ r_t \end{pmatrix} = \tilde{\Phi}_1 \begin{pmatrix} f_{t-1} \\ r_{t-1} \end{pmatrix} + \cdots + \tilde{\Phi}_p \begin{pmatrix} f_{t-p} \\ r_{t-p} \end{pmatrix} + \tilde{\varepsilon}_t^f \tag{3.22}$$

where $\tilde{\varepsilon}_t^f$ is i.i.d., $N(0, \tilde{\sum}^f)$. We interpret the factors as summarizing the information contained in a large dataset of macroeconomic, financial and commodities time-series. The FAVAR model is estimated by following a two-step procedure: (i) the PCA is used to extract the factors, and (ii) the parameters of the state equation are estimated with standard VAR methods.[11]

3.2.2.3 Impulse-Response Function Analysis

In this section, the purpose of our analysis consists in conducting an impulse response function (IRF) analysis in the FAVAR framework. Indeed, this econometric

---

[11] Standard initial conditions can be found in Koop (2003, [52]). This numerical approach is easier to implement than Markov Chain Monte Carlo Methods (MCMC). Note also that the PCA provides a solution only to the factor equation, without taking into account the dynamics of the factors (see Stock and Watson (2005, [67])).

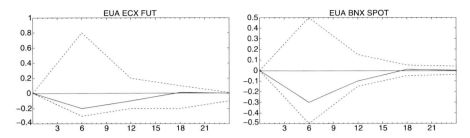

Fig. 3.8 IRF analysis of carbon futures (*left panel*) and spot (*right panel*) prices following a recessionary shock in the FAVAR model

strategy will allow us to understand how carbon markets react to an exogenous shock on the factors.

As in Bernanke et al. (2005, [11]), this task can be carried out by identifying 'slow-' vs. 'fast-moving' variables. The former type of variable does not respond contemporaneously to unanticipated shocks, and corresponds to output, production, consumption, employment, or interest rates. The latter type of variable adjusts immediately to economic shocks, and corresponds to asset prices. This identification scheme is necessary to interpret shocks coming from the macroeconomic, financial or commodity spheres.

By simulating a recessionary shock in the macroeconomic, financial and commodity environment, we obtain in Fig. 3.8 the corresponding impulse response functions for the carbon spot and futures markets. While bearing in mind that this stylized exercise corresponds only to a statistical artefact, such an exogenous shock can be thought from a modelling viewpoint as having an adverse effect on the economy comparable to the 2008 sub-primes crisis for instance.

Interestingly, the results show that the carbon price variables tend to respond *negatively* (in the range of -0.2 standard deviation) to an exogenous shock,[12] but with a delay (of three periods in our setting). Then, the effects of the shock die out, which is an indication of the stationarity of the FAVAR model. This result therefore tends to confirm the findings related to stock and bond markets, in a broader setting.

To sum up, compared to the previous section, the main conclusion remains unchanged: we can identify a link between carbon spot and futures markets on the one hand, and the macroeconomy on the other hand, but this link is rather weak compared to other asset classes. The result can be explained by the predominant institutional features of carbon markets as constructed 'environmental' markets. For instance, in October 2008, the main economic indicators were showing a sharp decline, in the midst of the 'credit-cruch'. However, the carbon price has been adjusting to this situation only from January 2009 onwards.

Finally, the interested reader may refer to Ludvigson and Ng [54, 55] and Zagaglia (2010, [81]) for further applications of factor models to, respectively, bond risk premia and the oil market.

---

[12]Hence, this result is conform to the relationship between carbon markets and the macroeconomy based on purely theoretical grounds.

In the next section, we investigate further how EUAs are expected to respond to the macroeconomy through variation in industrial production and the underlying business cycle.

3.3 Industrial Production

This section models the link between carbon prices and industrial production—coined as 'the carbon-macroeconomy relationship'—which unfolds as follows: as industrial production increases, associated CO_2 emissions increase and therefore more CO_2 allowances are needed by operators to cover their emissions. This economic logic results in carbon price increases, due to tighter constraints on the demand side of the market *ceteris paribus*.

To our best knowledge, only Alberola et al. (2008a, 2009) have opened this 'black box' by developing econometric analyses at the sector and country levels. They note that economic activity is perhaps the most obvious and least understood driver of CO_2 price changes. As economic growth leads to increased energy demand and higher industrial production in general, Alberola et al. [2, 3] show that carbon price changes react non only to energy prices forecast errors and extreme temperatures events, but also to industrial production in three sectors (combustion, paper, iron) and in four countries (Germany, Spain, Poland, UK) covered by the EU ETS.

In what follows, we develop a wide range of nonlinearity tests in order to understand the underlying dynamics of the EU industrial production and carbon prices. Then, we investigate the presence of threshold nonlinearities in each time series based on self-exciting threshold autoregression. Finally, we show how to analyze jointly industrial production and carbon prices based on smooth transition and Markov-switching autoregressive models.

Let us first present the data used.

3.3.1 Data

Figure 3.9 shows the two time series under consideration[13] in this section from January 2005 to July 2010. The source of the data is Eurostat and ECX. NBER business cycles reference dates are represented by gray vertical lines.

For industrial production, the EU 27 seasonally adjusted industrial production index is gathered in monthly frequency from Eurostat. For carbon prices, the European Union Allowance (EUA) Futures price is gathered in daily frequency from the European Climate Exchange (ECX). Then, the monthly EUA futures price is simply

---

[13]Note this data is not available for download for this chapter.

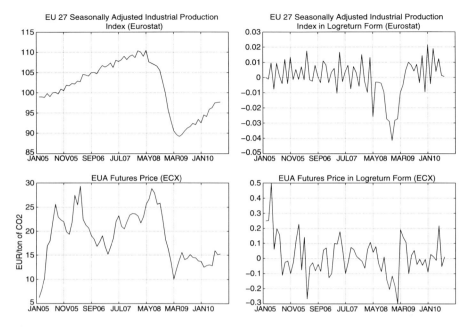

Fig. 3.9 EU 27 Industrial Production Index (*top*) and EUA Futures Price (*bottom*) in raw (*left panel*) and logreturn (*right panel*) forms from January 2005 to July 2010

computed as the average value of daily observations during a given month. Table 3.3 shows the descriptive statistics.[14]

Based on visual inspection, it appears that none of the time series under consideration is stationary in raw form. In the right panel of Fig. 3.9, however, the time series look somewhat stationary when transformed to logreturns. This first diagnostic is confirmed by usual unit root tests (ADF, PP, KPSS) in Table 3.4.[15]

Therefore, both variables are integrated of order 1 ($I(1)$).

---

[14]Std. Dev. is the standard deviation. JB stands for the Jarque Bera test. LB stands for the Ljung-Box test, whose *p*-values have been computed with a number of 20 lags (the values found are qualitatively similar with 10 or 15 lags). The same comments apply for the Engle ARCH test.

[15]For the ADF and PP tests, the null hypothesis is *EU27PRODINDRET* (*EUAFUTRET*) has a unit root (where *EU27PRODINDRET* (*EUAFUTRET*) stands for the EU27 Seasonally Adjusted Industrial Production Index in Logreturn form (the EUA Futures Price in Logreturn form)). For the ADF test, a lag length of 1 (0) is specified based on the Schwarz Information Criterion. For the PP test, a Bartlett kernel of bandwith 5 (1) is specified using the Newey-West procedure. For both tests, Model 1 (without trend nor intercept) is chosen. Test critical values at the 5% level are based on MacKinnon (1996). For the KPSS, the null hypothesis is *EU27PRODINDRET* (*EUAFUTRET*) is stationary. A Bartlett kernel of bandwidth 5 (3) is specified using the Newey-West procedure. Asymptotic critical values at the 5% level are based on KPSS (1992). Model 2 (with intercept) (Model 3 (with intercept and deterministic trend)) is chosen.

Table 3.3 Descriptive statistics

	EU27INDPROD	EU27INDPRODRET	EUAFUT	EUAFUTRET
Mean	101.3036	−0.000199	18.92626	0.013585
Median	102.1396	0.000295	19.06000	−0.004721
Maximum	110.4501	0.021329	29.33158	0.497638
Minimum	89.18148	−0.041477	6.225364	−0.299232
Std. Dev.	6.269672	0.012285	5.240281	0.131340
Skewness	−0.387235	−1.060101	−0.091494	0.637325
Kurtosis	1.960821	4.506348	2.371070	4.813195
JB	4.689154	18.60194	1.197730	13.50913
Prob. JB	0.095888	0.000091	0.549435	0.001166
LB test (p-value)	0.000001	0.000020	0.000001	0.413700
ARCH test (p-value)	0.002400	0.082600	0.070500	0.588600
Observations	67	66	67	66

Table 3.4 Unit root tests

EU27PRODINDRET	t-Statistic	Test critical values
Augmented Dickey-Fuller test statistic	−2.528185	−1.945987
Phillips-Perron test statistic	−6.382038	−1.945903
	LM-Stat.	Asymptotic critical values
Kwiatkowski-Phillips-Schmidt-Shin test statistic	0.166097	0.463000
EUAFUTRET	t-Statistic	Test critical values
Augmented Dickey-Fuller test statistic	−5.953323	−1.945903
Phillips-Perron test statistic	−5.892110	−1.945903
	LM-Stat.	Asymptotic critical values
Kwiatkowski-Phillips-Schmidt-Shin test statistic	0.108755	0.146000

In the right panel of Fig. 3.9, there seems to remain some instability in the transformed time series, more especially for the EU 27 Industrial Production Index in logreturn form (between May 2008 and March 2009). To investigate further this question, we have run Ordinary Least Squares-Cumulative Sum of Squares (OLS-CUSUM, Ploberger and Kramer ([62], 1992)) tests, which are based on cumulated sums of OLS residuals against a single-shift alternative.

As shown in Fig. 3.10, these results do not allow us to identify any kind of structural instability. Indeed, for both variables the empirical fluctuation processes stay

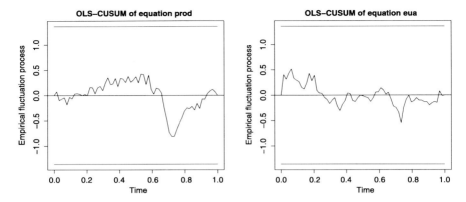

Fig. 3.10 OLS-CUSUM Test for EU27PRODINDRET (*left panel*) and EUAFUTRET (*right panel*)

safely within their bounds and do not seem to indicate the presence of structural breaks in the data. Next, we introduce the nonlinearity tests.

3.3.2 Nonlinearity Tests

This section proposes several nonlinearity tests to analyze the characteristics of the data. It is generally recognized that macroeconomic and financial time series (such as equity markets) are characterized by the occurrence of shocks. Because of the existence of these shocks, a linear model of the underlying returns may provide a misleading specification of market movements. If the dynamic propagation of shocks differs from the 'usual' behavior of the time series, then a model that relies on a linear propagation mechanism will necessarily be incorrect (Bradley and Jansen (2004, [17])).

In response to these limitations, nonlinear time series models generally display a rich dynamical structure (Franses and Van Dijk (2003, [45]), Cryer and Chan (2008, [30])). Nonlinear time series analysis was first developed during the late 1970s, prompted by the need for modeling the nonlinear dynamics shown by real data. Except for cases with well-developed theory accounting for the underlying mechanism of an observed time series, the nonlinear data mechanism is generally unknown. Thus, a fundamental problem of empirical nonlinear time series analysis concerns the choice of a general nonlinear class of models.

Several tests have been proposed for assessing the need for nonlinear modeling in time series analysis (Tong (1990, [75]), Granger and Teräsvirta (1993, [47])). Since there exists by nature an infinity of nonlinear models, we restrict our attention to some of these tests studied by Keenan (1985, [51]), Tsay (1986, [77]) and Brock et al. [6, 8].

3.3.2.1 Keenan (1985)

Let $Y_1, \ldots, Y_n$ denote the observations. Keenan (1985, [51])'s test may be derived by considering the following model:

$$Y_t = \theta_0 + \phi_1 Y_{t-1} + \cdots + \phi_m Y_{t-m} + \exp\left\{ \eta \left(\sum_{j=1}^{m} \phi_j Y_{t-j} \right)^2 \right\} + \varepsilon_t \qquad (3.23)$$

where the ϕ's are autoregressive parameters, m is some pre-specified positive integer, and ε_t is the error term which is assumed to be i.i.d. with zero mean and finite variance. If $\eta = 0$, the exponential term becomes 1 and can be absorbed into the intercept term, so that the preceding model becomes an $AR(m)$ model. On the other hand, for nonzero η, the preceding model is nonlinear. Therefore, Keenan (1985)'s test is the same as the F-test for testing whether or not $\eta = 0$ by using the following test statistic:

$$\hat{F} = \frac{\eta^2(n - 2m - 2)}{RSS - \eta^2} \qquad (3.24)$$

with RSS the residual sum of squares. Under the null hypothesis of linearity, the test statistic $\hat{F}$ is approximately distributed as an F-distribution with degrees of freedom 1 and $(n - 2m - 2)$. A more formal approach is facilitated by the Lagrange multiplier test (see Tong (1990, [75])).

3.3.2.2 Tsay (1986)

Keenan (1985, [51])'s test is powerful only for detecting nonlinearity in the form of the square of the approximating linear conditional mean function (Cryer and Chan (2008, [30])). Tsay (1986, [77]) extended this approach by considering more general nonlinear alternatives:

$$Y_t = \theta_0 + \phi_1 Y_{t-1} + \cdots + \phi_m Y_{t-m} \qquad (3.25)$$

$$+ \delta_{1,1} Y_{t-1}^2 + \delta_{1,2} Y_{t-1} Y_{t-2} + \cdots + \delta_{1,m} Y_{t-1} Y_{t-m} \qquad (3.26)$$

$$+ \delta_{2,2} Y_{t-2}^2 + \delta_{2,3} Y_{t-2} Y_{t-3} + \cdots + \delta_{2,m} Y_{t-2} Y_{t-m} + \cdots \qquad (3.27)$$

$$+ \delta_{m-1,m-1} Y_{t-m+1}^2 + \delta_{m-1,m} Y_{t-m+1} Y_{t-m} + \delta_{m,m} Y_{t-m}^2 + \varepsilon_t \qquad (3.28)$$

where, using $\exp(x) \approx 1 + x$, we see that the nonlinear model is approximately a quadratic AR model. The coefficients of the quadratic terms are now unconstrained, and Tsay (1986, [77])'s test is equivalent to testing whether or not all the $m(m+1)/2$ coefficients $\delta_{i,j}$ are zero. Again, this can be carried out by an F-test that all $\delta_{i,j}$'s are zero in the preceding equation. For a rigorous derivation of Tsay (1986, [77])'s test as a Lagrange multiplier test, see Tong (1990, [75]).

3.3.2.3 Brock et al. (1987, 1996)

Brock et al. (1987, [6])'s test (published later in Brock, Dechert and Scheinkman (1996, [8]), henceforth BDS) was originally designed to test for the null hypothesis of independent and identical distribution (i.i.d.) for the purpose of detecting non-random chaotic dynamics.[16] However, some studies have shown that this test has power against a wide range of linear and nonlinear alternatives (see for example Brock et al. (1991, [7])). If the null hypothesis cannot be rejected, then the original linear model cannot be rejected. If the null hypothesis is rejected, the fitted linear model is mis-specified, and in this sense, it may also be viewed as a test for nonlinearity.

The main concept behind the BDS test is the correlation integral, which is a measure of the frequency with which temporal patterns are repeated in the data. Consider a time series x_t for $t = 1, 2, \ldots, T$ and define its m-history as $x_t^m = (x_t, x_{t-1}, \ldots, x_{t-m+1})$. The correlation integral at embedding dimension m can be estimated by:

$$C_{m,\delta} = \frac{2}{T_m(T_m - 1)} \sum_{m \leq s < t \leq T} \sum I(x_t^m, x_s^m, \delta) \qquad (3.29)$$

where $T_m = T - m + 1$ and $I(x_t^m, x_s^m, \delta)$ is an indicator function which is equal to one if $|x_{t-i} - x_{s-i}| < \delta$ for $i = 0, 1, \ldots, m - 1$ and zero otherwise. Intuitively, the correlation integral estimates the probability that any two m-dimensional points are within a distance of δ of each other. If x_t are i.i.d., this probability should be equal to $C_{1,\delta}^m = Pr(|x_t - x_s| < \delta)^m$ in the limiting case. Brock et al. (1996) define the BDS statistic as follows:

$$V_{m,\delta} = \sqrt{T} \frac{C_{m,\delta} - C_{1,\delta}^m}{s_{m,\delta}} \qquad (3.30)$$

where $s_{m,\delta}$ is the standard deviation of $\sqrt{T}(C_{m,\delta} - C_{1,\delta}^m)$, and may be estimated consistently as documented by Brock et al. (1987, [6]). Under fairly moderate regularity conditions, the BDS statistic converges in distribution to $\mathcal{N}(0, 1)$, so that the null hypothesis of i.i.d. is rejected at the 5% significance level whenever $|V_{m,\delta}| > 1.96$.

3.3.2.4 Results

Keenan (1985, [51]) and Tsay (1986, [77]) nonlinearity tests results are shown in Tables 3.5 and 3.6.

In Table 3.5, to carry out the nonlinearity tests, we have to specify the working autoregressive order m. Under the null hypothesis that the process is linear, the order can be specified by minimizing some information criterion, for example the Akaike Information Criterion (AIC). For *EU27INDPRODRET*, $m = 6$ based on the AIC.

---

[16]Loosely speaking, a time series is said to be 'chaotic' if it follows a nonlinear deterministic process, but looks random.

Table 3.5 Keenan (1985) and Tsay (1986) nonlinearity tests results for *EU27INDPRODRET*

	Test Statistic	*p*-value	Order
Keenan (1985)	1.8238	0.1827	6
Tsay (1986)	3.0470	0.0022	6

Table 3.6 Keenan (1985) and Tsay (1986) nonlinearity tests results for *EUAFUTRET*

	Test Statistic	*p*-value	Order
Keenan (1985)	0.2277	0.6348	1
Tsay (1986)	0.1218	0.7282	1

Table 3.7 Brock et al. (1987, 1996) test results for independence and identical distribution for *EU27INDPRODRET*

Embedding dimension =	[2]	[3]		
δ for close points	[0.0061]	[0.0123]	[0.0184]	[0.0246]
Test Statistics	[0.0061]	[0.0123]	[0.0184]	[0.0246]
[2]	6.5196	6.6728	6.2597	6.7664
[3]	9.3868	7.2437	6.2892	6.8695
p-value	[0.0061]	[0.0123]	[0.0184]	[0.0246]
[2]	0	0	0	0
[3]	0	0	0	0

Interestingly, we obtain contradictory results. The Keenan (1985, [51]) test does not reject linearity (with a *p*-value being 0.1827), whereas the Tsay (1986, [77]) test clearly rejects linearity (with a *p*-value equal to 0.0022).

In Table 3.6, we consider the time series of *EUAFUTRET*. The working AR order is found to be 1. Both the Keenan (1985, [51]) test and the Tsay (1986, [77]) test fail to reject linearity, with *p*-values being 0.6348 and 0.7282, respectively.

Brock et al. [6, 8] test results for independence and identical distribution are presented in Tables 3.7 and 3.8.

In Table 3.7, the null hypothesis is that *EU27INDPRODRET* is independently and identically distributed. The default values of $\delta = (0.5, 1.0, 1.5, 2.0)$ used in the test are converted back to the units of the original data, and the null hypothesis that the data is i.i.d. is rejected for most combinations of *m* and δ at conventional significance levels. Therefore, the results from the BDS test suggest that there may be a nonlinear structure in the data. The same comments apply for *EUAFUTRET* in Table 3.8.

As summarized in Table 3.9, we obtain mixed empirical evidence concerning the presence of nonlinearities in the data. The Tsay and BDS tests indicate non-linearities in the structure of *EU27INDPRODRET*, while the Keenan test does not

Table 3.8 Brock et al. (1987, 1996) test results for independence and identical distribution for *EUAFUTRET*

Embedding dimension =	[2]	[3]		
δ for close points	[0.0657]	[0.1313]	[0.1970]	[0.2627]
Test Statistics	[0.0657]	[0.1313]	[0.1970]	[0.2627]
[2]	6.5231	5.3256	4.8749	4.6857
[3]	10.8570	6.6217	5.5321	5.5028
p-value	[0.0657]	[0.1313]	[0.0197]	[0.2627]
[2]	0	0	0	0
[3]	0	0	0	0

Table 3.9 Summary of Nonlinearity Tests Results

Evidence of Nonlinearity	Keenan (1985)	Tsay (1986)	Brock et al. (1987, 1996)
EU27INDPRODRET	No	Yes	Yes
EUAFUTRET	No	No	Yes

reject linearity. For *EUAFUTRET*, the Keenan and Tsay tests fail to reject linearity, while the BDS test suggests a nonlinear structure. The fact that we find contradictory results between the Keenan and Tsay tests for *EU27INDPRODRET* is indeed puzzling. Based on the visual inspection of Fig. 3.9, we have concluded that there remains maybe some instability in the time series of *EU27INDPRODRET* (due to the recessionary shock of the sub-primes crisis). Therefore, this question related to the existence of several regimes needs to be further investigated in the next section.

3.3.3 Self-exciting Threshold Autoregressive Models

In this section, we consider the univariate modelling of the time series. Indeed, an understanding of the univariate dynamics of the industrial production and carbon price returns constitutes an important starting point for a posterior analysis of their joint distribution.

Threshold models constitute the simplest class of nonlinear model. Their usefulness in nonlinear time series analysis was well-documented by the seminal work of Tong [73–75] and Tong and Lim (1980, [76]), resulting in an extensive literature of ongoing theoretical innovations and applications in various fields. There exist many variants of the threshold model. Here, we focus on the two-regime self-exciting threshold autoregressive (SETAR) model, for which the switching between the two linear submodels depends solely on the position of the threshold variable (see Tong (1990, [75]), Chan (1993, [22]), Chan and Tsay (1998, [25]), and the references therein). For the SETAR model, the threshold variable is a certain lagged value of the process itself: hence the adjective self-exciting. Consider the following model (Cryer and Chan (2008, [30])):

$$Y_t = \begin{cases} \phi_{1,0} + \phi_{1,1}Y_{t-1} + \cdots + \phi_{1,p_1}Y_{t-p_1} + \sigma_1 e_t, & \text{if } Y_{t-d} \leq r \\ \phi_{2,0} + \phi_{2,1}Y_{t-1} + \cdots + \phi_{2,p_2}Y_{t-p_2} + \sigma_2 e_t, & \text{if } Y_{t-d} > r \end{cases} \qquad (3.31)$$

where the ϕ's are autoregressive parameters, σ's are noise standard deviations, r is the threshold parameter, and $\{e_t\}$ is a sequence of independent and identically distributed random variables with zero mean and unit variance. Note that the autoregressive orders p_1 and p_2 of the two submodels need not be identical, and the delay parameter d may be larger than the maximum autoregressive orders. The model defined by Eq. (3.31) is denoted as the SETAR$(2; p_1, p_2)$ model with delay d. The SETAR model is ergodic and hence asymptotically stationary if $|\phi_{1,1}| + \cdots + |\phi_{1,p}| < 1$ and $|\phi_{2,1}| + \cdots + |\phi_{2,p}| < 1$ (see Chan and Tong (1985, [23])).

In what follows, we first test for threshold nonlinearity and unit root. Second, we estimate SETAR models separately for the EU industrial production and carbon futures prices to gain more knowledge about the behavior of each time series.

3.3.3.1 Testing for Threshold Nonlinearity and Unit Root

While Keenan (1985, [51])'s and Tsay (1986, [77])'s tests for nonlinearity are designed for detecting quadratic nonlinearity, they may not be sensitive to *threshold* nonlinearity. Hence, we introduce below a likelihood ratio (LR) test with the threshold model as the specific alternative. The null hypothesis is an AR(p) model versus the alternative hypothesis of a two-regime SETAR model of order p and with constant noise variance, i.e. $\sigma_1 = \sigma_2 = \sigma$. With these assumptions, Eq. (3.31) may be rewritten as (Cryer and Chan (2008, [30])):

$$Y_t = \phi_{1,0} + \phi_{1,1}Y_{t-1} + \cdots + \phi_{1,p}Y_{t-p} + \{\phi_{2,0} + \phi_{2,1}Y_{t-1} + \cdots$$
$$+ \phi_{2,p}Y_{t-p}\}I(Y_{t-d} > r) + \sigma e_t \qquad (3.32)$$

where $I(.)$ is an indicator variable that equals 1 if and only if the enclosed expression is true. In this specification, the coefficient $\phi_{2,0}$ represents the change in the intercept in the upper regime relative to that of the lower regime, and similarly interpreted are the $\phi_{2,1}, \ldots, \phi_{2,p}$ coefficients. The null hypothesis states that $\phi_{2,0} = \phi_{2,1} = \cdots = \phi_{2,p} = 0$. While the delay may be theoretically larger than the autoregressive order, this is seldom the case in practice. Hence, it is assumed that $d \leq p$ and assuming the validity of linearity, the large-sample distribution of the test does not depend on d. In practice, the test is carried out with fixed p and d. The Likelihood Ratio (LR) test statistic writes:

$$T_n = (n - p) \log \left\{ \frac{\hat{\sigma}^2(H_0)}{\hat{\sigma}^2(H_1)} \right\} \qquad (3.33)$$

where $n - p$ is the effective sample size, $\hat{\sigma}^2(H_0)$ is the maximum likelihood estimator of the noise variance from the linear AR(p) fit, and $\hat{\sigma}^2(H_1)$ from the SETAR

Table 3.10 LR test for threshold nonlinearity for *EU27INDPRODRET* with $p = 6$

d	1	2	3	4	5
Test Statistic	46.570	43.209	64.870	25.273	58.466
p-value	0.000	0.000	0.000	0.012	0.000

Table 3.11 LR test for threshold nonlinearity for *EUAFUTRET* with $p = 1$

d	1	2	3	4	5
Test Statistic	1.133	1.597	4.494	3.037	7.250
p-value	0.116	0.311	0.341	0.297	0.192

Note: d is the delay argument

fit with the threshold searched over some finite interval. Under the null hypothesis, the (nuisance) parameter r is absent. Hence, the sampling distribution of the likelihood ratio test under H_0 is no longer approximately χ^2 with p degrees of freedom. Instead, it has a nonstandard sampling distribution. Chan (1991, [21]) derived an approximation method for computing the p-values of the test that is highly accurate for small p-values. The test depends on the interval over which the threshold parameter is searched. Typically, the interval is defined to be from the $a \times 100$th percentile to the $b \times 100$th percentile of $\{Y_t\}$. The choice of a and b must ensure that there are adequate data falling into each of the two regimes for fitting the linear submodels.[17]

LR tests results for threshold nonlinearity, assuming normally distributed innovations, are shown in Tables 3.10 and 3.11.

Recall that in Tables 3.5 and 3.6, the optimal autoregressive orders based on the AIC are equal to $p = 6$ and $p = 1$, respectively, for *EU27INDPRODRET* and *EUAFUTRET*. Therefore, we have plugged in these values to conduct the LR test for threshold nonlinearity in Tables 3.10 and 3.11.

Setting $a = 0.25$ and $b = 0.75$ for *EU27INDPRODRET*,[18] we have tried the LR test for threshold nonlinearity with different delays $d = 1, \ldots, 5$, resulting in the test statistics shown in Table 3.10. All the tests above have p-values less than 1%, suggesting that the data-generating mechanism is highly nonlinear. Notice that the test statistic attains the largest value when $d = 3$. Next, consider Table 3.11: with $a = 0.25$, $b = 0.75$, and $1 \leq d \leq 5$, the test statistics and their p-values do not exhibit any evidence of threshold nonlinearity, with the delay likely to be 5.

---

[17]This restriction is necessary because if the true model is linear, the threshold parameter is undefined, in which case an unrestricted search may result in the threshold estimator being close to the minimum or maximum data values, making the large-sample approximation ineffective (Cryer and Chan (2008, [30])).

[18]Note that repeating the test with $a = 0.1$ and $b = 0.9$ yields identical results. This comment applies in the remainder of the chapter.

Going back to the comments of Table 3.9, LR tests for threshold nonlinearity tend to confirm the presence of a nonlinear structure in the time series of *EU27INDPRODRET*. For *EUAFUTRET* however, these results taken together rather lead to conclude against the evidence of nonlinearities in the data (despite the result of the BDS test).

Next, we consider tests that allow for the joint consideration of nonlinearity (threshold) and nonstationarity (unit roots). As pointed out by Pippenger and Goering [60, 61] and Dufrenot and Mignon (2002, [33]), there is a need to develop methods to detect stationarity (or cointegration) when the underlying process follows a threshold model. Caner and Hansen (2001, [20]) have derived such a test, which has been applied recently by Basci and Caner (2005, [5]):

$$\Delta y_t = \theta_1' x_{t-1} + e_t \quad \text{if } |y_{t-1} - y_{t-m-1}| < \lambda \tag{3.34}$$

$$\Delta y_t = \theta_2' x_{t-1} + e_t \quad \text{if } |y_{t-1} - y_{t-m-1}| \geq \lambda \tag{3.35}$$

where y_t is the selected time series, $x_{t-1} = (y_{t-1}, 1, \Delta y_{t-1}, \ldots, \Delta y_{t-k})'$ for $t = 1, 2, \ldots, T$. e_t is the i.i.d. error term, m represents the delay parameter, and $1 \leq m \leq k$. The threshold variable is the absolute value of $y_{t-1} - y_{t-m-1}$. The threshold value λ is unknown and takes the value in the compact interval $\lambda \in \Lambda = [\lambda_1, \lambda_2]$ where these values are picked according to $P(|y_{t-1} - y_{t-m-1}| \leq \lambda_1) = 0.15$, $P(|y_{t-1} - y_{t-m-1}| \leq \lambda_2) = 0.85$. Next, we decompose the coefficients:

$$\theta_1 = \begin{pmatrix} \rho_1 \\ \beta_1 \\ \alpha_1 \end{pmatrix}, \qquad \theta_2 = \begin{pmatrix} \rho_2 \\ \beta_2 \\ \alpha_2 \end{pmatrix}$$

where ρ_1 and ρ_2 are scalar, β_1 and β_2 have the same dimension as y_t, α_1 and α_2 are $k \times 1$ vectors. (ρ_1, ρ_2) represent the slope coefficients on y_{t-1}, (β_1, β_2) are the slopes on the deterministic components, and (α_1, α_2) are the slope coefficients on $(\Delta y_{t-1}, \ldots, \Delta y_{t-k})$ in the two regimes (see the underlying assumptions for this model in Caner and Hansen (2001, [20])). Equation (3.34) may be rewritten as:

$$\Delta y_t = \theta_1' x_{t-1} 1_{\{|y_{t-1} - y_{t-m-1}| < \lambda\}} + \theta_2' x_{t-1} 1_{\{|y_{t-1} - y_{t-m-1}| \geq \lambda\}} + e_t \tag{3.36}$$

with $1_{\{.\}}$ the indicator function. Under the null of unit root, we have $H_0 : \rho_1 = \rho_2 = 0$. As emphasized by Basci and Caner (2005, [5]), there are two interesting alternatives at hand. First, if the time series follows a stationary threshold autoregressive pattern, the alternative of interest is $H_1 : \rho_1 < 0, \rho_2 < 0$. Second, there is the case of partial unit root:

$$H_2 : \begin{cases} \rho_1 < 0 & \text{and} \quad \rho_2 = 0 \\ \rho_1 = 0 & \text{and} \quad \rho_2 < 0 \end{cases} \tag{3.37}$$

If H_2 holds, then the time series is nonstationary, but we do not deal with a classic unit root. The test statistics for testing H_0 vs. H_1 and H_0 vs. H_2 are given by Caner and Hansen (2001, [20]). Since H_1 is one-sided, we test $H_0 : \rho_1 = \rho_2 = 0$ with a simple one-sided Wald as test statistic:

$$R_{1T} = t_1^2 1_{\{\hat{\rho}_1 < 0\}} + t_2^2 1_{\{\hat{\rho}_2 < 0\}} \tag{3.38}$$

Table 3.12 Bootstrap p-values of threshold unit root tests (Caner and Hansen 2001)

Variable	R_{1T}	t_1	t_2
EU27INDPROD	0.002*	0.274	0.011*
EUAFUT	0.002*	0.014*	0.322

Note: R_{1T}, t_1, t_2 are unit root tests described in Caner and Hansen (2001). R_{1T} is for testing H_0 vs. H_1. t_1, t_2 are tests for H_0 vs. H_2. Under their respective columns, we report the bootstrap p-values. m is estimated by minimizing the Sum of Squared Errors (SSE). Bootstrap p-values are calculated from 10,000 replications. *represents significance at the 10% level

with t_1, t_2 the t-ratios for respectively $\hat{\rho}_1$ and $\hat{\rho}_2$. In order to test H_0 vs. H_2, we use the negative of the t statistics $-t_1$, $-t_2$. For both unit root tests, Caner and Hansen (2001, [20]) have obtained the limit distributions, and the critical values are tabulated in Table 3 of their article. Even though the unit root tests have the asymptotic bound distribution, we may benefit from a bootstrap in finite samples.

Now, let us use the one-sided Wald test (R_{1T}) and t_1, t_2 tests for unit roots to determine whether the regimes identified above are nonstationary or not. We use the bootstrap procedure described in Sect. 5.3 of Caner and Hansen (2001, [20]), since this seems to work better in finite samples. Hence, the bootstrap p-values are reported for the one-sided Wald and the t-ratio tests in Table 3.12.

Both R_{1T} statistics are significant at the 10% level, with bootstrap p-values of 0.002. Therefore, the one-sided Wald test (unit root vs. two-regime stationary nonlinear model) is rejected for both time series.

For *EU27INDPROD*, the individual t ratios t_1 and t_2 have bootstrap p-values of 0.274, and 0.011, respectively. The rejection is due to the second regime, where the p-value for the t test is statistically significant.

Turning to *EUAFUT*, the t_1 and t_2 tests have bootstrap p-values of 0.014, and 0.322, respectively. The rejection is due to the first regime, where the p-value for the t test is statistically significant.

Together, these test statistics indicate that $\rho_1 = 0$ and $\rho_2 < 0$ for *EU27INDPROD* and $\rho_1 < 0$ and $\rho_2 = 0$ for *EUAFUT*. Thus, we may conclude that both time series are partially stationary threshold processes.

3.3.3.2 Univariate Results

Estimation of the SETAR Model   Cryer and Chan (2008, [30]) detail the estimation methodologies of the SETAR models, based on log-likelihood functions or conditional least squares. In practice, the AR orders in the two regimes need not be identical or known. Thus, an efficient estimation procedure that also estimates the orders is essential. Recall that for linear ARMA models, the AR orders can be estimated by minimizing the AIC. For fixed r and d, the SETAR model is essentially fitting two AR models of orders p_1 and p_2, respectively, so that the AIC becomes:

$$AIC(p_1, p_2, r, d) = -2 \log(maximum\ likelihood, r, d) + 2(p_1 + p_2 + 2) \quad (3.39)$$

Table 3.13 AIC of the SETAR Models Fitted to *EU27INDPRODRET*

d	AIC	$\hat{r}$	$\hat{p}_1$	$\hat{p}_2$
1	−405.80	−0.004	6	5
2	−409.30	−0.004	6	3
3	−427.70	−0.002	4	1
4	−399.60	−0.004	6	3
5	−409.60	−0.001	6	3

Note: d is the delay argument. $\hat{r}$ is the estimated threshold parameter. $\hat{p}_1$ and $\hat{p}_2$ are, respectively, the estimated orders of the two AR models fitted by the SETAR model

Table 3.14 AIC of the SETAR Models Fitted to *EUAFUTRET*

d	AIC	$\hat{r}$	$\hat{p}_1$	$\hat{p}_2$
1	−102.50	0.052	3	3
2	−97.79	0.011	0	0
3	−106.00	0.059	1	4
4	−101.40	0.011	1	2
5	−100.20	−0.007	0	1

Note: d is the delay argument. $\hat{r}$ is the estimated threshold parameter. $\hat{p}_1$ and $\hat{p}_2$ are, respectively, the estimated orders of the two AR models fitted by the SETAR model

where the number of parameters (excluding r, d, σ_1, and σ_2) equals $p_1 + p_2 + 2$. We estimate the parameters by minimizing the AIC, subject to the constraint that the threshold parameter be searched over some interval that guarantees any regimes have adequate data for estimation. By using this method with the search of a threshold between the 10th and 90th percentiles, Tables 3.13 and 3.14 display the nominal AIC value for $1 \le d \le 5$.

Although the maximum autoregressive order is set to 6, minimizing the AIC for *EU27INDPRODRET* yields to select the order 4 for the lower regime and the order 1 for the upper regime (Table 3.13). For *EUAFUTRET*, we select the order 1 for the lower regime and the order 4 for the upper regime by using the same methodology (Table 3.14). For both time series, the nominal AIC is smallest when $d = 3$, so we set the delay to 3.

Next, we summarize in Tables 3.15 and 3.16 the corresponding models fit. The submodel in each regime is estimated by ordinary least squares using the data falling in that regime. The 'unbiased' noise variance $\tilde{\sigma}_i^2$ of the ith regime relates to its maximum likelihood counterpart by the formula:

$$\tilde{\sigma}_i^2 = \frac{n_i}{n_i - p_i - 1}\hat{\sigma}_i^2 \tag{3.40}$$

where p_i is the autoregressive order of the ith submodel. Moreover, $(n_i - p_i - 1)\tilde{\sigma}_i^2/\sigma_i^2$ is approximately distributed as χ^2 with $n_i - p_i - 1$ degrees of freedom.

Table 3.15 Fitted SETAR$(2, 4, 1)$ Model with $d = 3$ for *EU27INDPRODRET*

	Estimate	Std. Error	t-statistic	p-value
$\hat{d}$	3			
$\hat{r}$	-0.002			
Lower Regime ($n_1 = 27$)				
$\hat{\phi}_{1,0}$	-0.013	0.003	-4.688	0.001
$\hat{\phi}_{1,1}$	0.263	0.150	1.755	0.093
$\hat{\phi}_{1,2}$	1.098	0.161	6.842	0.001
$\hat{\phi}_{1,3}$	-0.808	0.222	-3.642	0.001
$\hat{\phi}_{1,4}$	-0.358	0.138	-2.605	0.016
$\tilde{\sigma}_1^2$	0.008			
Upper Regime ($n_2 = 35$)				
$\hat{\phi}_{2,0}$	0.007	0.001	6.878	0.001
$\hat{\phi}_{2,1}$	-0.545	0.967	-5.636	0.001
$\tilde{\sigma}_2^2$	0.005			

Note: d is the delay argument. n_i with $i = 1, 2$ is the number of observations in each sub-regime. The model estimated is defined in Eq. (3.31)

Table 3.15 displays the estimation results of the SETAR$(2, 4, 1)$ model with $d = 3$ fitted to *EU27INDPRODRET*. The threshold estimate is -0.002 for lag 3 values. The lower regime contains 27 observations. The upper regime is composed of 35 observations. In general, a threshold estimate that is too close to the minimum or the maximum observation may be unreliable due to small sample size in one of the regimes which, fortunately, is not the case here. Notice that, in both regimes, all coefficient estimates are statistically significant. Therefore, this model provides a good fit to the data.

Table 3.16 displays the estimation results of the SETAR$(2, 1, 4)$ model with $d = 3$ fitted to *EUAFUTRET*. The threshold estimate is 0.059 for lag 3 values. The lower regime contains 43 observations, while the upper regime has only 17 observations. Notice that the coefficients of the intercept in the lower regime and of lag 2 in the upper regime are not statistically significant. Therefore, the estimation of the SETAR model is less satisfactory for *EUAFUTRET* (compared to *EU27INDPRODRET*). This is in line with our previous comments regarding the rejection of nonlinearities in this time series.

An interesting question concerns the interpretation of the two regimes. One way to explore the nature of the regimes is to identify which data value falls in which regime in the time series plots of Fig. 3.11. In the top panel, the data falling in the lower regime for *EU27INDPROD* (i.e., whose lag 3 values are less than -0.002) are displayed as solid circles, whereas those in the upper regime are shown as open circles. Similarly, in the bottom panel, data falling in the lower regime for *EUAFUT* (i.e., whose lag 3 values are less than 0.059) are displayed as solid circles. This

Table 3.16 Fitted SETAR$(2, 1, 4)$ Model with $d = 3$ for *EUAFUTRET*

	Estimate	Std. Error	t-statistic	p-value
$\hat{d}$	3			
$\hat{r}$	0.059			
Lower Regime ($n_1 = 43$)				
$\hat{\phi}_{1,0}$	-0.025	-0.003	0.017	0.884
$\hat{\phi}_{1,1}$	0.306	0.306	0.0147	0.043
$\tilde{\sigma}_1^2$	0.109			
Upper Regime ($n_1 = 17$)				
$\hat{\phi}_{2,0}$	0.134	0.043	3.121	0.009
$\hat{\phi}_{2,1}$	-0.535	0.164	-3.271	0.007
$\hat{\phi}_{2,2}$	0.077	0.119	0.646	0.530
$\hat{\phi}_{2,3}$	-1.113	0.310	-3.595	0.004
$\hat{\phi}_{2,4}$	-0.187	0.096	-1.946	0.076
$\tilde{\sigma}_2^2$	0.063			

Note: d is the delay argument. n_i with $i = 1, 2$ is the number of observations in each sub-regime. The model estimated is defined in Eq. (3.31)

figure reveals that the estimated lower regime (solid circles) corresponds mainly to the recession period for *EU27INDPROD*, and to periods of decreasing prices for *EUAFUT*. Conversely, the estimated upper regime (open circles) corresponds mainly to periods before/after the recession for *EU27INDPROD*, and to periods of increasing prices for *EUAFUT*. Interestingly, note that from January 2009 onwards the carbon market is characterized by a lower regime, indicating a delayed adjustment to the financial crisis.

Model Diagnostics   Here, we discuss some formal statistical approaches to model diagnostics via residual analysis (Cryer and Chan (2008, [30])). The raw residuals $\hat{\varepsilon}_t$ are given by:

$$\hat{\varepsilon}_t = Y_t - \{\hat{\phi}_{1,0} + \hat{\phi}_{1,1} Y_{t-1} + \cdots + \hat{\phi}_{1,p} Y_{t-p}\} I (Y_{t-\hat{d}} \leq \hat{r})$$
$$- \{\hat{\phi}_{2,0} + \hat{\phi}_{2,1} Y_{t-1} + \cdots + \hat{\phi}_{2,p} Y_{t-p}\} I (Y_{t-\hat{d}} > \hat{r}) \qquad (3.41)$$

The standardized residuals $\hat{e}_t$ are obtained by:

$$\hat{e}_t = \frac{\hat{\varepsilon}_t}{\hat{\sigma}_1 I (Y_{t-\hat{d}} \leq \hat{r}) + \hat{\sigma}_2 I (Y_{t-\hat{d}} > \hat{r})} \qquad (3.42)$$

As in the linear case, the time series plot of the standardized residuals should look random, as they should be approximately independent and identically distributed if the SETAR model is correctly specified. Hence, we need to look for the presence of outliers and any systematic pattern in such a plot, in which case it may provide a clue for specifying a more appropriate model. The independence assumption of

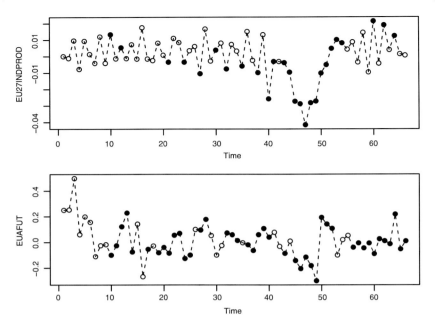

Fig. 3.11 Thresholds Estimated by the SETAR Models for the EU 27 Industrial Production Index (*top panel*) and EUA Futures Price (*bottom panel*). Note: *Solid* (*open*) *circles* indicate data in the *lower* (*upper*) regime of a fitted threshold autoregressive model

the standardized errors can be checked by examining the ACF of the standardized residuals. Nonconstant variance may be checked by examining the sample ACF of the absolute standardized residuals.

Besides, we consider the generalization of the portmanteau test based on some overall measure of the magnitude of the residual autocorrelations. The dependence of the residuals necessitates the employment of a quadratic form of the residual autocorrelations:

$$B_m = n_{eff} \sum_{i=1}^{m} \sum_{j=1}^{m} q_{i,j} \hat{\rho}_i \hat{\rho}_j \qquad (3.43)$$

where $n_{eff} = n - \max(p_1, p_2, d)$ is the effective sample size, $\hat{\rho}_i$ is the ith lag sample autocorrelation of the standardized residuals, and $q_{i,j}$ some model-dependent constants given in Cryer and Chan (2008, [30]). If the true model is a SETAR model, $\hat{\rho}_i$ are likely close to zero and so is B_m, but B_m tends to be large if the model specification is incorrect. The quadratic form is designed so that B_m is approximately distributed as χ^2 with m degrees of freedom. In practice, the p-value of B_m may be plotted against m over a range of m values to provide a more comprehensive assessment of the independence assumption on the standardized errors.

Model diagnostics are shown in Figs. 3.12 and 3.13. The top panel of Fig. 3.12 represents the time series plot of the standardized residuals of the SETAR(2, 4, 1) model with $d = 3$ fitted to *EU27INDPRODRET*. Except for some possible outlier, the plot shows no particular pattern (as the standardized residuals are scattered around zero). The middle panel is the ACF plot of the standardized residuals. The confidence band is based on the simple $1.96/\sqrt{n}$ rule, and should be regarded as a rough guide on the significance of the residual ACF. No lags in the residual autocorrelation are found to be significant. The bottom panel reports the p-values of the more rigorous portmanteau test. The p-values are found to be very large for all m. As no p-value is found to be significant (i.e. we accept the null hypothesis of no autocorrelation in the residuals), we may infer that the SETAR(2, 4, 1) model is well-specified. Similar diagnostics for the SETAR(2, 1, 4) model with $d = 3$ for *EUAFUTRET* model are shown in Fig. 3.13.

In short, this section has been devoted to the univariate time series analysis of the EU industrial production and carbon futures returns based on SETAR models. The fitted SETAR models seem to provide a better fit to *EU27INDPRODRET* than to *EUAFUTRET*, owing to the nonlinear properties of the respective time series. Nevertheless, we have gained useful knowledge concerning the identification of thresholds in the data. First, the time series of the EU industrial production is characterized by two regimes before and after the recessionary period, which correspond to the starting/trough dates of the financial crisis in the US. Second, the time series of the EU carbon futures prices is characterized by two regimes depending on increasing or decreasing price trends. From January 2009 onwards, we identify a delayed adjustment of carbon markets (and more generally of commodity markets, see Chevallier (2011, [27])) to the global economic recession. Besides, we have checked that the residuals of the estimated models are not autocorrelated.

Let us now extend our analysis to two kinds of bivariate threshold models, in order to analyze both time series jointly.

3.3.4 Comparing Smooth Transition and Markov-Switching Autoregressive Models

It is a virtually unquestioned assumption that fluctuations in the level of economic activity are a key determinant of the level of carbon prices: as industrial production increases, associated CO_2 emissions increase and therefore more CO_2 allowances are needed by operators to cover their emissions.

We aim here at extending the analysis of the carbon-macroeconomy relationship based on two kinds of time-series models. First, threshold regime-switching happens when the threshold variable crosses a certain value. Although that type of model may capture many nonlinear features usually observed in economic time series, it seems counter-intuitive to suggest that the regime switch is abrupt or discontinuous. Hence, the main interest behind smooth transition autoregressive (STAR) models is to assume that the regime switch happens gradually in a 'smooth' fashion.[19] The observations may switch smoothly between regimes, with one regime

---

[19]The discontinuity of the thresholds is replaced by a smooth transition function (typically the logistic or exponential functions, see Van Dijk et al. (2002, [79]) for an exhaustive review of STAR models).

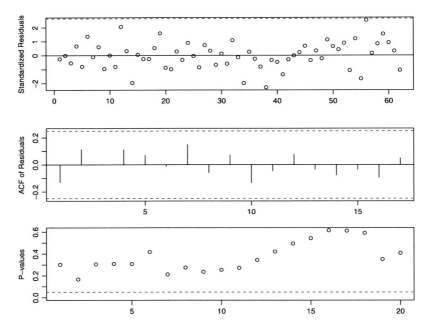

Fig. 3.12 Model Diagnostics of the SETAR(2, 4, 1) Model with $d = 3$ for *EU27INDPRODRET*

having more impacts in some times and other regimes having more impacts in other times. Another interpretation is that STAR models actually allow for a 'continuum' of regimes, each associated with a different value of the transition function. This type of model has attracted a lot of attention in previous literature. For instance, Teräsvirta and Anderson (1992, [72]) characterize nonlinearities in business cycles, such as production and unemployment, using smooth transition autoregressive models.

Second, Markov-switching models have been widely used in economics and finance since Hamilton (1989, [48]) introduced them to estimate regime- or state-dependent variables. They have also been utilized to capture volatility in financial markets. Cai (1994, [19]) and Hamilton and Susmel (1994, [50]) introduce Markov-switching models to estimate high- and low-volatility regimes in financial data. Regimes constructed in this way are an important instrument for interpreting business cycles. They constitute an optimal inference on the latent state of the economy, whereby probabilities are assigned to the unobserved regimes 'expansion' and 'contraction' conditional on the available information set. Clearly, such an approach is useful when a series is thought to undergo shifts from one type of behaviour to another and back again, but where the 'forcing variable' that causes the regime shifts is unobservable.

Therefore, these two kinds of models appear complementary, as they allow us to explore 'smooth' vs. 'abrupt' changes in the carbon-macroeconomy relationship. Therefore, it appears worthy of interest to compare the results obtained for these

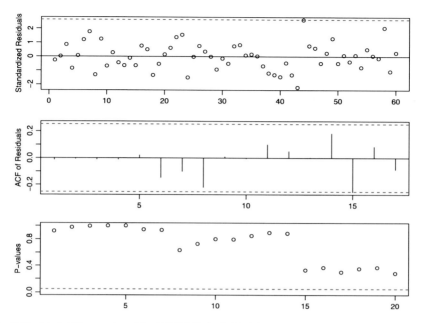

Fig. 3.13   Model Diagnostics of the SETAR$(2, 1, 4)$ Model with $d = 3$ for *EUAFUTRET*

two econometric specifications. Such an econometric exercise has been performed by Deschamps (2008, [31]) for instance, in the context of US unemployment.

3.3.4.1 Model Specification

In what follows, we set *EUAFUTRET* as the endogenous variable y_t, and *EU27IND PRODRET* as part of the exogenous variables x_t (with other variables such as autoregressive and lagged components).

Smooth Transition AutoRegression   A criticism of the SETAR models developed is that the conditional mean equation is not continuous. The thresholds are the discontinuity points of the conditional mean function. In response to this criticism, Smooth Threshold AutoRegressive Models (STAR) models have been proposed by Chan and Tong (1986, [24]) and Teräsvirta (1994, [69]) (among others, see Van Dijk et al. (2002, [79]) for a review).[20]

According to Teräsvirta (2004, [71]), the standard STAR model with a *logistic* transition function (henceforth, LSTAR) has the form:

---

[20]See also Bradley and Jansen (2004, [17]) for an application of STAR models to stock returns and industrial production.

$$y_t = \phi' \mathbf{z_t} + \theta' \mathbf{z_t}\, G(\gamma, c, s_t) + u_t, \quad u_t \sim \text{i.i.d.}(0, \sigma^2) \tag{3.44}$$

$$G(\gamma, c, s_t) = \left(1 + \exp\left\{-\gamma \prod_{k=1}^{K}(s_t - c_k)\right\}\right)^{-1}, \quad \gamma > 0 \tag{3.45}$$

where $\mathbf{z_t} = (\mathbf{w_t'}, \mathbf{x_t'})'$ is an $((m+1) \times 1)$ vector of explanatory variables with $\mathbf{w_t'} = (1, y_{t-1}, \ldots, y_{t-p})'$ and $\mathbf{x_t'} = (x_{1t}, \ldots, x_{kt})'$. ϕ and θ are the parameter vectors of the linear and the nonlinear parts, respectively. The transition function $G(\gamma, c, s_t)$ depends on the transition variable s_t, the slope parameter γ and the vector of location parameters c. The most common choices for K are $K = 1$ (*LSTAR1*) and $K = 2$ (*LSTAR2*). The transition variable s_t can be part of $\mathbf{z_t}$ or it can be another variable, like for example $s_t = t$ (trend). The following steps occur when working with STAR models.

Specifying the AR Part   First, we need to select the linear AR model to start from. The maximum lag order for y_t and $\mathbf{x_t}$ determines the number of lags to include.[21]

Testing Linearity   Second, we may test whether there is a nonlinearity of the STAR type in the model (Luukkonen et al. (1988, [57])). It also helps to determine the transition variable, and whether *LSTAR1* or *LSTAR2* should be used. The following auxiliary regression is applied if s_t is an element of $\mathbf{z_t}$:

$$y_t = \beta_0' \mathbf{z_t} + \sum_{j=1}^{3} \beta_j' \tilde{\mathbf{z}}_\mathbf{t} s_t^j + u_t^* \tag{3.46}$$

with $\mathbf{z_t} = (1, \tilde{\mathbf{z}}_\mathbf{t})'$. The null hypothesis of linearity is $H_0 : \beta_1 = \beta_2 = \beta_3 = 0$. This linear restriction is checked by applying a F-test.

Choosing the Type of Model   Third, once linearity has been rejected, one has to choose whether an LSTAR1 or an LSTAR2 model should be specified. The choice can be based on the following test sequence: $H_{04} : \beta_3 = 0$, $H_{03} : \beta_2 = 0|\beta_3 = 0$, $H_{02} : \beta_1 = 0|\beta_2 = \beta_3 = 0$. This test is based on the same auxiliary regression as the linearity test.[22]

Finding Initial Values   Fourth, to find the initial values, the parameters of the STAR model are estimated by a nonlinear optimization routine. The gridsearch creates a linear grid in c and a log-linear grid in γ. For each value of γ and c, the residual sum of squares is computed. The values that correspond to the minimum of that sum are taken as starting values.[23]

---

[21]Note that the transition variable s_t must be part of the lags of these variables if it is not a trend.

[22]Note this task may also be performed by looking at the information criteria, or at the residual sum of squares.

[23]Note also that in order to make γ scale-free, it is divided by $\hat{\sigma}_s^K$, the Kth power of the sample standard deviation of the transition variable.

Estimation of Parameters   Fifth, once good starting values have been found, the unknown parameters c, γ, θ, ϕ may be estimated by using a form of the Newton-Raphson algorithm to maximize the conditional maximum likelihood function (see Lütkepohl and Krätzig (2004, [56]) for more details).

Testing the STAR Model   Sixth, the quality of the estimated nonlinear model should be checked against misspecification (see Van Dijk et al. (2002, [79]) for a detailed exposition). The tests for STAR models are generalizations of the corresponding tests for misspecification in linear models.[24] The test of no error autocorrelation is a special case of the general test described by Godfrey (1988, [46]). It has been discussed in its application to STAR models by Teräsvirta (1998, [70]). The procedure is to regress the estimated residuals $\tilde{u}_t$ on lagged residuals $\tilde{u}_{t-1}, \tilde{u}_{t-q}$ and the partial derivatives of the log-likelihood function with respect to the parameters of the model. The test statistic is then:

$$F_{LM} = \frac{\{(SSR_0 - SSR_1)/q\}}{\{SSR_1/(T - n - q)\}} \tag{3.47}$$

with n the number of parameters in the model, SSR_0 the sum of squared residuals of the STAR model and SSR_1 the sum of squared residuals from the auxiliary regression.

Next, we need to check whether there is remaining nonlinearity in the model. To that purpose, the test of no additive nonlinearity assumes that the type of the remaining nonlinearity is again of the STAR type. The alternative can be defined as:

$$y_t = \phi' \mathbf{z_t} + \theta' \mathbf{z_t} \, G(\gamma_1, c_1, s_{1t}) + \psi' \mathbf{z_t} \, H(\gamma_2, c_2, s_{2t}) + u_t \tag{3.48}$$

with H another transition function and $u_t \sim$ i.i.d.$(0, \sigma^2)$. To test this alternative, the following auxiliary model is used:

$$y_t = \beta_0' \mathbf{z_t} + \theta' \mathbf{z_t} \, G(\gamma_1, c_1, s_{1t}) + \sum_{j=1}^{3} \beta_j' \tilde{\mathbf{z}}_t s_{2t}^j + u_t^* \tag{3.49}$$

The test is performed by regressing $\tilde{u}_t$ on $(\tilde{\mathbf{z}}_t', s_{2t}, \tilde{z}_t' \, s_{2t}^2, \tilde{z}_t' \, s_{2t}^3)'$ and the partial derivatives of the log-likelihood function with respect to the parameters of the model. The null hypothesis of no remaining nonlinearity is that $\beta_1 = \beta_2 = \beta_3 = 0$. The choice of s_{2t} may be a subset of available variables in $\mathbf{z_t}$, or it may be s_{1t}. The resulting F-statistics are given as for the test on linearity.

Finally, the test of parameter constancy verifies the null hypothesis of constant parameters against smooth continuous change in parameters. The alternative may be written as follows:

$$y_t = \phi(t)' \mathbf{z_t} + \theta(t)' \mathbf{z_t} \, G(\gamma, c, s_t) + u_t, \quad u_t \sim \text{i.i.d.}(0, \sigma^2) \tag{3.50}$$

---

[24]Note that the traditional ARCH LM test for the presence of heteroskedasticity and the Jarque-Bera test for normality may also be developed for STAR models.

with $\phi(t) = \phi + \lambda_\phi H_\phi(\gamma_\phi, c_\phi, t^*)$, $\theta(t) = \theta + \lambda_\theta H_\theta(\gamma_\theta, c_\theta, t^*)$, and $t^* = \frac{t}{T}$. The null hypothesis of no change in parameters is $\gamma_\phi = \gamma_\theta = 0$. The parameters γ and c are assumed to be constant. The following nonlinear auxiliary regression is used:

$$y_t = \beta_0' \mathbf{z_t} + \sum_{j=1}^{3} \beta_j' \mathbf{z_t}(t^*)^j + \sum_{j=1}^{3} \beta_{j+3}' \mathbf{z_t}(t^*)^j \ G(\gamma, c, s_t) + u_t^* \tag{3.51}$$

The corresponding F-test results are given for three alternative transition functions:

$$H(\gamma, c, t^*) = \left(1 + \exp\left\{-\gamma \prod_{k=1}^{K}(t^* - c_k)\right\}\right)^{-1} - \frac{1}{2}, \quad \gamma > 0 \tag{3.52}$$

with $K = 1, 2, 3$ respectively, and assuming that $\gamma_\phi = \gamma_\theta$.

Markov-Switching Alternatively, the presence of multiple regimes may be acknowledged using a popular multivariate model introduced by Hamilton (1989, [48]), where parameters are made dependent on a hidden state process ruled by a Markov chain: such a model is called Markov-switching model. Let us consider the following specification to study the carbon-macroeconomy relationship:

$$y_t = \mu_{S_t} + \beta_{1,S_t} x_{1t} + \beta_{2,S_t} x_{2t} + \beta_{3,S_t} x_{3t} + \varepsilon_t, \quad \varepsilon_t \sim \mathcal{N}(0, \sigma_{S_t}^2) \tag{3.53}$$

with $S_t = 1, \ldots, K$ the state at time t where K is the number of states, $\sigma_{S_t}^2$ the error variance at state S_t, β_{i,S_t} the coefficient for the explanatory variable i at state S_t, and ε_t the residual vector which is assumed to follow a Gaussian distribution. The regime-generating process is an ergodic Markov chain with a finite number of states defined by the transition probabilities:

$$p_{ij} = \text{Prob}(S_{t+1} = j | s_t = i), \quad \sum_{j=1}^{K} p_{ij} = 1 \quad \forall i, j \in \{1, \ldots, K\} \tag{3.54}$$

where the optimal inference about the unobserved state variable S_t takes the form of a probability.

Similarly to the STAR model, the endogenous variable y_t is *EUAFUTRET*. In the Markov-switching specification, we include three explanatory variables with $i = \{EUAFUTRET(-1), EU27INDPRODRET, EU27INDPRODRET(-1)\}$. β_{1,S_t} captures the influence of the AR(1) process, β_{2,S_t} the influence of *EU27INDPRODRET*, and β_{3,S_t} the influence of *EU27INDPRODRET(-1)*.

By setting $S = [1\ 1\ 1\ 1]$, both the β_{i,S_t} coefficients and the model's variance are switching according to the transition probabilities. Typically, in order to track the 'boom-bust' economic cycle, we set the number of states K equal to 2. Therefore, state $K = 1$ represents the 'high growth' phase, whereas state $K = 2$ characterizes the 'low growth' phase (for more details, see Hamilton (2008, [49]) and references therein). When $K = 1$, the growth of the endogenous variable is given by the population parameter μ_1, whereas when $K = 2$, the growth rate is μ_2.

Table 3.17 p-values of the linearity test against STAR modelling

Hypothesis	Transition variable		
	$EUAFUTRET(-1)$	$EU27INDPRODRET$	$EU27INDPRODRET(-1)$
H_0	0.00023	0.00576	0.00381
H_{04}	0.22380	0.06847	0.06985
H_{03}	0.69885	0.21148	0.38651
H_{02}	0.00001	0.00789	0.00011
Model chosed	$LSTAR1$	$LSTAR1$	$LSTAR1$

Table 3.18   Grid search results for the starting values of γ and c

Parameter	Transition variable		
	$EUAFUTRET(-1)$	$EU27INDPRODRET$	$EU27INDPRODRET(-1)$
γ	9.0186	10.0000	10.0000
c	−0.0285	−0.1069	−0.1069
SSR	0.0047	0.0051	0.0054

The model is estimated based on Gaussian maximum likelihood with $S_t = 1, 2$. **P** is the transition matrix which controls the probability of a switch from state $K = 1$ to state $K = 2$:

$$\mathbf{P} = \begin{bmatrix} p_{11} & p_{21} \\ p_{12} & p_{22} \end{bmatrix}$$

The sum of each column in **P** is equal to 1, since they represent the full probabilities of the process for each state.

3.3.4.2   Results

Smooth Transition AutoRegression   Table 3.17 presents the results of the linearity test against STAR modelling.[25]

For all the transition variables considered, the p-values are remarkably small. In each case, because the p-value of the test for H_{03} is much larger than the ones corresponding to testing H_{04} and H_{02}, the choice of the LSTAR1 model is quite clear. Table 3.18 presents the grid search results for the starting values of γ and c by minimizing the Sum of Squared Residuals (SSR) of the STAR model.

Tables 3.19 and 3.20 present the estimation results of the STAR models.[26]

---

[25]The (-1) term into parentheses means that the variable is lagged one period.

[26]Note for the transition variable $EUAFUTRET(-1)$, the STAR model has not been estimated owing to near singularity of the moment matrix.

Table 3.19 Estimation results of the *LSTAR*1 model with the transition variable *EU27INDPRODRET*

Variable Linear Part	Estimate	Std. Dev.	t-stat.	p-value
Constant	−0.03558	0.0127	−2.8024	0.0070
EUAFUTRET(−1)	−0.10713	0.4073	0.2631	0.7935
EU27INDPRODRET	0.12510	0.0595	−2.1018	0.0402
EU27INDPRODRET(−1)	−0.12631	0.0581	2.1734	0.0341
Nonlinear Part				
Constant	0.03855	0.0128	3.0196	0.0038
EUAFUTRET(−1)	0.37668	0.4263	−0.8836	0.3808
EU27INDPRODRET	−0.11834	0.0605	1.9556	0.0556
EU27INDPRODRET(−1)	0.12370	0.0593	−2.0860	0.0416
γ	55.49749	61.6420	0.9003	0.3719
c	−0.11187	0.0056	−20.0185	0.0000
Diagnostic tests				
AIC	0.3453			
SC	0.0107			
HQ	0.2133			
Adjusted R^2	0.5809			
Variance of transition variable	0.0166			
Std. Dev. of transition variable	0.1290			
Variance of residuals	0.0001			
Std. Dev. of residuals	0.0087			

In Table 3.17, it has been seen that using *EU27INDPRODRET* as transition variable results in the smallest *p*-value. Hence, Table 3.19 provides the estimation results where *EU27INDPRODRET* is selected as being the appropriate transition variable in a possible STAR model for the carbon-macroeconomy relationship. In both linear and nonlinear components, we may remark that *EU27INDPRODRET* and *EU27INDPRODRET*(−1) have statistically significant coefficient estimates.[27] Therefore, we are able to identify that some effects in the carbon-macroeconomy relationship channel from industrial production to carbon price changes. As for the signs of the coefficients obtained, we may tentatively attempt to explain that contem-

---

[27]Note that the estimate of γ (the slope parameter) is not significant. Its large standard deviation estimate reflects the numerical difficulties in estimating γ accurately when it is large, and the transition function is thus close to a step function (for a more detailed discussion of this phenomenon, see for example Granger and Teräsvirta (1993, [47]) or Teräsvirta [69, 70]).

Table 3.20 Estimation results of the *LSTAR*1 model with the transition variable *EU27INDPRODRET*(−1)

Variable Linear Part	Estimate	Std. Dev.	*t*-stat.	*p*-value
Constant	−0.03096	0.0166	1.8609	0.0681
EUAFUTRET(−1)	−0.72039	0.2184	3.2987	0.0017
EU27INDPRODRET	0.08077	0.0327	2.4693	0.0167
EU27INDPRODRET(−1)	−0.14137	0.0715	1.9785	0.0529
Nonlinear Part				
Constant	0.02866	0.0168	−1.7092	0.0931
EUAFUTRET(−1)	1.04911	0.2668	−3.9316	0.0002
EU27INDPRODRET	−0.07794	0.0346	−2.2526	0.0283
EU27INDPRODRET(−1)	0.14237	0.0722	−1.9718	0.0537
γ	29.57175	22.1172	1.3370	0.1867
c	−0.10670	0.0058	−18.4580	0.0000
Diagnostic tests				
AIC	0.1371			
SC	0.8025			
HQ	0.0051			
Adjusted R^2	0.4839			
Variance of transition variable	0.0175			
Std. Dev. of transition variable	0.1324			
Variance of residuals	0.0001			
Std. Dev. of residuals	0.0097			

poraneous changes in the industrial production index impact *negatively* carbon price changes (i.e. the decrease in industrial production precedes the decrease in carbon prices), while changes in the industrial production index lagged one period impact *positively* carbon price changes (i.e. the uptake in economic activity encourages the carbon price to go up). Besides, as seen in Fig. 3.14 where each dot corresponds to an observation, the transition function as a function of the observations has a steep slope. The results obtained are qualitatively similar when *EU27INDPRODRET*(−1) is selected as the transition variable (Table 3.20).

According to Table 3.21, the nonlinearities indicated by the results in Table 3.17 have been adequately modeled (i.e. the *p*-values are large).

Results of tests of parameter constancy may be found in Table 3.22.[28]

They show, by and large, that the STAR models estimated have constant parameters. Given again the large *p*-values obtained (in the last column of Table 3.22),

---

[28]NA stands for 'Not Available' when the test encounters a matrix inversion problem. *df* stands for degree of freedom.

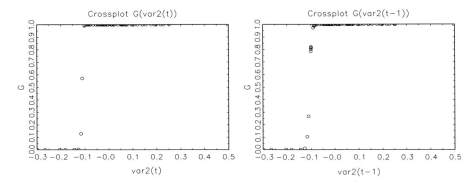

Fig. 3.14 Transition function of the STAR model as a function of the observations for the transition variables *EU27INDPRODRET* (*left panel*), and *EU27INDPRODRET*(−1) (*right panel*)

Table 3.21 *p*-values of the test of no additive nonlinearity in STAR models

Hypothesis	Transition variable	
	EU27INDPRODRET	*EU27INDPRODRET*(−1)
H_0	0.31482	0.89850
H_{04}	0.49260	0.91746
H_{03}	0.62651	0.48974
H_{02}	0.87790	0.72642

Table 3.22 *p*-values of parameter constancy tests in STAR models

Transition Variable	*EU27INDPRODRET*			
Transition Function	*F*-value	df1	df2	*p*-value
H_1	0.9202	8.0000	45.0000	0.5090
H_2	1.0498	16.0000	37.0000	0.4324
H_3	NA	NA	NA	NA
Transition Variable	*EU27INDPRODRET*(−1)			
Transition Function	*F*-value	df1	df2	*p*-value
H_1	0.6478	8.0000	45.0000	0.7335
H_2	0.8122	16.0000	37.0000	0.6640
H_3	1.2496	24.0000	29.0000	0.2812

the STAR models estimated therefore pass all parameter constancy tests, and this issue is not deemed worthy of further investigation. Other diagnostic tests do not indicate misspecification: the Jarque-Bera normality test and the ARCH-LM test have a *p*-value equal to, respectively, 0.18 and 0.72 (0.01 and 0.32) when

Table 3.23 Test of no error autocorrelation in STAR models up to lag 8

Transition Variable	EU27INDPRODRET			
Lag	F-value	df1	df2	p-value
1	0.0109	1	53	0.9171
2	5.2987	2	51	0.8100
3	3.9422	3	49	0.1350
4	3.0792	4	47	0.2480
5	2.4220	5	45	0.5000
6	2.5939	6	43	0.3110
7	2.1392	7	41	0.6060
8	1.7813	8	39	0.1105
Transition Variable	EU27INDPRODRET(−1)			
Lag	F-value	df1	df2	p-value
1	1.5433	1	53	0.2196
2	3.8507	2	51	0.2770
3	5.1381	3	49	0.3600
4	4.1786	4	47	0.5600
5	3.6815	5	45	0.7100
6	3.7627	6	43	0.4300
7	3.7249	7	41	0.3300
8	3.3488	8	39	0.5200

EU27INDPRODRET (*EU27INDPRODRET*(−1)) is selected as the transition variable.

Finally, Table 3.23 indicates that the residuals of the STAR models are not autocorrelated.

Given the information set, our STAR model appears wholly adequate to represent the carbon-macroeconomy relationship. This conclusion however does not exclude the possibility that some other nonlinear models would not fit the data better, as investigated in the next section.

Markov-Switching Table 3.24 reports the estimation results for the univariate Markov-switching model.[29]

An examination of the coefficients of the two means (μ_{S_t}), which are all statistically significant, shows the presence of switches in growth between the two regimes.

---

[29]Standard errors are in parentheses. ***, **, * denote respectively statistical significance at the 1%, 5% and 10% levels.

Table 3.24 Estimation results of the univariate Markov-switching model

Log-likelihood	51.35
μ (Regime 1)	0.0066***
	(0.0014)
μ (Regime 2)	−0.0041***
	(0.0148)

Equation for *EUAFUTRET*	*EUAFUTRET*(−1)	*EU27INDPRODRET*	*EU27INDPRODRET*(−1)
β_i (Regime 1)	0.2191*	2.0289*	1.1821
	(0.1298)	(1.0601)	(1.1343)
β_i (Regime 2)	0.2238	2.6879	−1.2769
	(0.3258)	(6.3797)	5.4435
Standard error (Regime 1)	0.0092		
Standard error (Regime 2)	0.0163		

Transition Probabilities Matrix	Regime 1	Regime 2
Regime 1	0.9312***	0.3213
	(0.1206)	(0.2130)
Regime 2	0.0721	0.6800***
	(0.0668)	(0.0982)

Regime Properties	Prob.	Duration
Regime 1	0.6833	14.12
Regime 2	0.3167	3.16

In Regime 1 (expansion), output growth per month is equal to 0.66% on average, while in Regime 2 (recession) the average growth rate amounts to −0.41%. The AR(1) is significant at the 10% level. We also find that *EU27INDPRODRET* has a contemporaneous and positive impact on *EUAFUTRET* at the 10% level during Regime 1 ('high growth'). Other coefficient estimates are not statistically significant.

The bottom lines of Table 3.24 report the matrix of transition probabilities for the latent variable S_t (standard error in parentheses). During an expansionary phase, the series are most likely to remain in Regime 1 (with an estimated probability equal to 93.12%). On the contrary, the probability that the series switch from Regime 1 to Regime 2 is lower (equal to 07.21%). Once the economy finds itself in a depression, the probability that it will be in a depression the following month is estimated to be 68.00%. Finally, if the economy is in Regime 2 (recessionary

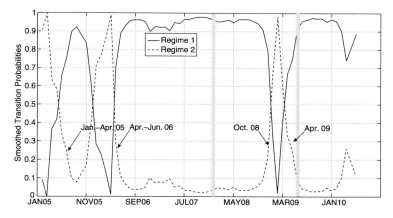

Fig. 3.15 Smoothed transition probabilities estimated from the univariate Markov-switching model

phase), the probability that it will change directly to a growth regime is equal to 32.13%.

Let us now have a look at the average duration for each regime. While Regime 2 is assumed to last 3.16 months on average, the average duration of an expansionary phase is equal to 14.12 months. Therefore, the transition probabilities associated with each regime point out that the first regime is more persistent. The ergodic probabilities imply that the economy would spend about 68% of the time spanned by our sample of data in the first regime (i.e. expansion). In contrast, Regime 2 has an ergodic probability of about 32%. Finally, Regime 2 exhibits a relatively higher standard error (0.0163) than Regime 1 (0.0092), which is conform to the view that periods of recession are less stable than periods of expansion.

Figures 3.15 and 3.16 show, respectively, the associated smoothed and regime transition probabilities.[30] Switches from one regime to another are especially perceptible during January-April 2005, April-June 2006, October 2008 and April 2009. As for the first two dates, one may cautiously attempt to relate them to early market developments of the EU ETS. From January to April 2005, market agents had heterogeneous anticipations with regard to the actual level of carbon prices in a context of sustained EU economic growth. In April 2006, carbon prices were characterized by a strong downward adjustment due to a situation of 'over-allocation' compared to verified CO_2 emissions (Ellerman and Buchner (2008, [34])). This situation of high price volatility lasted until the end of June 2006 (Alberola et al. (2008, [1])). Concerning the third date, October 2008 corresponds to the first high growth-low growth

---

[30]The regime (smoothed) probability at time t is the probability that state t will operate at t, conditional on information available up to $t-1$ (conditional on all information in the sample). Regime 1 is 'expansion'. Regime 2 is 'contraction'. NBER business cycles reference dates are represented by gray vertical lines.

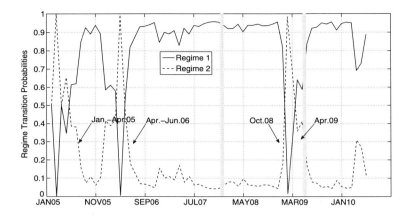

Fig. 3.16 Regime transition probabilities estimated from the univariate Markov-switching model

switch in the carbon-macroeconomy relationship. Indeed, EU industrial production had been falling since July 2007. However, the carbon market seemed to adjust to this situation only in October 2008 when most operators were looking to sell allowances in exchange of cash (Chevallier (2009, [26])). Finally, the recent period is characterized by a delayed adjustment of most commodity markets (among which the carbon market) to the global recessionary shock (Caballero et al. (2008, [18]), Tang and Xiong (2009, [68])).

To sum up, this section provides the first empirical analysis of the carbon-macroeconomy relationship through smooth transition autoregression (STAR) and Markov-switching models from January 2005 to July 2010. Let us recall the economic rationale behind these two variables' relationship. First, economic activity fosters high demand for industrial production goods. In turn, companies falling under the regulation of the EU ETS need to produce more, and emit more CO_2 emissions in order to meet consumers' demand. This yields to a greater demand for CO_2 allowances to cover industrial emissions, and ultimately to carbon price increases.

On the one hand, when the EU industrial production index is selected as being the appropriate transition variable in the STAR model, we are able to identify that contemporaneous (lagged) changes in the industrial production index impact negatively (positively) EUA carbon price changes. Given the information set, STAR models thus appear to be a reasonable specification to capture the underlying nonlinearities. On the other hand, industrial production is found to impact positively (at the 10% level) carbon prices during expansion periods. Univariate Markov-switching modelling allows us to get a more precise understanding as to when the switches occur between high- and low-growth phases, i.e. during January-April 2005, April-June 2006, October 2008 and April 2009. These events have been cautiously related to the underlying macroeconomic situation and actual market developments in the EU ETS.

The key policy implications may be summarized as follows: (i) there is a link between macroeconomic activity and carbon price changes, (ii) this link seems to channel more precisely through the effects of industrial production (and associated CO_2 emissions) on carbon prices, and (iii) STAR and univariate Markov-switching models appear adequate to capture this link, while presenting complementary results. The Markov regime-switching model is also applied to the interactions between EUAs and CERs in the Appendix of Chap. 4. Finally, note that other econometric applications of threshold vector error-correction and Markov-switching VAR models applied to the carbon-macroeconomy relationship can be found in Chevallier (2011, [28, 29]).

References

1. Alberola E, Chevallier J, Cheze B (2008) Price drivers and structural breaks in European carbon prices 2005-07. Energy Policy 36:787–797
2. Alberola E, Chevallier J, Cheze B (2008) The EU emissions trading scheme: the effects of industrial production and CO_2 emissions on European carbon prices. Int Econ 116:93–125
3. Alberola E, Chevallier J, Cheze B (2009) Emissions compliances and carbon prices under the EU ETS: a country specific analysis of industrial sectors. J Policy Model 31:446–462
4. Bailey W, Chan KC (1993) Macroeconomic influences and the variability of the commodity futures basis. J Finance 48:555–573
5. Basci E, Caner M (2005) Are real exchange rates nonlinear or nonstationary? Evidence from a new threshold unit root test. Stud Nonlinear Dyn Econom 9:1–19
6. Brock W, Dechert WD, Scheinkman JA (1987) A test for independence based on the correlation dimension. Working paper, Department of Economics, University of Wisconsin, Madison, USA
7. Brock WA, Hsieh DA, LeBaron B (1991) Nonlinear dynamics, chaos, and instability: statistical theory and economic evidence. MIT Press, Cambridge
8. Brock WA, Dechert WD, Scheinkman JA, LeBaron B (1996) A test for independence based on the correlation dimension. Econom Rev 15:197–235
9. Benz E, Trück S (2009) Modeling the price dynamics of CO_2 emission allowances. Energy Econ 31:4–15
10. Bera AK, Higgins ML (1993) ARCH models: properties, estimation and testing. J Econ Surv 7:305–366
11. Bernanke B, Boivin J, Eliasz P (2005) Measuring monetary policy: a factor augmented vector autoregressive (FAVAR) approach. Q J Econ 120:387–422
12. Bessembinder H, Chan KC (1992) Time-varying risk premia and forecastable returns in futures markets. J Financ Econ 32:169–193
13. Berndt E, Hall B, Hall R, Hausman J (1974) Estimation and inference in nonlinear structural models. Ann Econ Soc Meas 3:653–665
14. Bollerslev T (1986) Generalized autoregressive conditional heteroskedasticity. J Econom 31:307–327
15. Bollerslev T, Chou RY, Kroner KF (1992) ARCH modeling in finance: a review of the theory and empirical evidence. J Econom 52:5–59
16. Bollerslev T, Engle RF, Nelson DB (1994) ARCH models. Handb Econom 4:2959–3038
17. Bradley MD, Jansen DW (2004) Forecasting with a nonlinear dynamic model of stock returns and industrial production. Int J Forecast 20:321–342
18. Caballero R, Farhi E, Gourinchas P (2008) Financial crash, commodity prices and global imbalances. Brookings Pap Econ Act 2:1–55

19. Cai J (1994) A Markov model of unconditional variance in ARCH. J Bus Econ Stat 12:309–316
20. Caner M, Hansen BE (2001) Threshold autoregression with a unit root. Econometrica 69:1555–1596
21. Chan KS (1991) Percentage points of likelihood ratio tests for threshold autoregression. J R Stat Soc B 53:691–696
22. Chan KS (1993) Consistency and limiting distribution of the least squares estimator of a continuous autoregressive model. Ann Stat 21:520–533
23. Chan KS, Tong H (1985) On the use of the deterministic Lyapunov function for the ergodicity of stochastic difference equations. Adv Appl Probab 17:666–678
24. Chan KS, Tong H (1986) On estimating thresholds in autoregressive models. J Time Ser Anal 7:179–190
25. Chan KS, Tsay RS (1998) Limiting properties of the conditional least squares estimator of a continuous TAR model. Biometrika 85:413–426
26. Chevallier J (2009) Carbon futures and macroeconomic risk factors: a view from the EU ETS. Energy Econ 31:614–625
27. Chevallier J (2011) Macroeconomics, finance, commodities: interactions with carbon markets in a data-rich model. Econ Model 28:557–567
28. Chevallier J (2011) Evaluating the carbon-macroeconomy relationship: Evidence from threshold vector error-correction and Markov-switching VAR models. Econ Model (forthcoming). doi:10.1016/j.econmod.2011.08.003
29. Chevallier J (2011) A model of carbon price interactions with macroeconomic and energy dynamics. Energy Econ (forthcoming). doi:10.1016/j.eneco.2011.07.012
30. Cryer JD, Chan KS (2008) Time series analysis with applications in R, 2nd edn. Springer texts in statistics. Springer, New York
31. Deschamps PJ (2008) Comparing smooth transition and Markov-switching autoregressive models of US unemployment. J Appl Econom 23:435–462
32. Ding Z, Granger CWJ, Engle RF (1993) A long memory property of stock market returns and a new model. J Empir Finance 1:83–106
33. Dufrenot G, Mignon V (2002) Recent developments in nonlinear cointegration with applications to macroeconomics and finance. Springer, New York
34. Ellerman AD, Buchner BK (2008) Over-allocation or abatement? A preliminary analysis of the EU ETS based on the 2005-06 emissions data. Environ Resour Econ 41:267–287
35. Engle RF (1982) Autoregressive conditional heteroskedasticity with estimates of variance of United Kingdom inflation. Econometrica 50:987–1008
36. Engle RF (2001) GARCH 101: the use of ARCH/GARCH models in applied econometrics. J Econ Perspect 15:157–168
37. Engle RF (2003) Risk and volatility: econometric models and financial practice. Nobel Lecture, December 2003
38. Engle RF, Lilien DM, Robins RP (1987) Estimating time varying risk premia in the term structure: the ARCH-M model. Econometrica 55:391–408
39. Engle RF, Patton AJ (2001) What good is a volatility model? Quant Finance 1:237–245
40. Fama EF, French KR (1987) Dividend yields and expected stock returns. J Financ Econ 22:3–25
41. Fama EF, French KR (1989) Business conditions and expected returns on stocks and bonds. J Financ Econ 25:23–49
42. Fama EF, French KR (1992) The cross-section of expected stock returns. J Finance 47:427–465
43. Fama EF, French KR (1993) Common risk factors in the returns on stocks and bonds. J Financ Econ 33:3–56
44. Fleming J, Ostdiek B (1999) The impact of energy derivatives on the crude oil market. Energy Econ 21:135–167
45. Franses PH, Van Dijk D (2003) Nonlinear time series models in empirical finance, 2nd edn. Cambridge University Press, Cambridge

46. Godfrey L (1988) Misspecification tests in econometrics. Cambridge University Press, Cambridge
47. Granger CWJ, Teräsvirta T (1993) Modelling nonlinear economic relationships. Oxford University Press, Oxford
48. Hamilton JD (1989) A new approach to the economic analysis of nonstationary time series and the business cycle. Econometrica 57:357–384
49. Hamilton JD (2008) Regime-switching models. In: Durlauf SN, Blume LE (eds) The new Palgrave dictionary of economics, 2nd edn. Palgrave Macmillan, Basingstoke, pp 1–15
50. Hamilton JD, Susmel R (1994) Autoregressive conditional heteroskedasticity and changes in regime. J Econom 64:307–333
51. Keenan DM (1985) A Tukey nonadditivity-type test for time series nonlinearity. Biometrika 72:39–44
52. Koop G (2003) Bayesian econometrics. Wiley, New York
53. Li WK, Ling S, McAleer M (2002) Recent theoretical results for time series models with GARCH errors. J Econ Surv 16:245–269
54. Ludvigson SC, Ng S (2007) The empirical risk-return tradeoff: a factor analysis approach. J Financ Econ 83:171–222
55. Ludvigson SC, Ng S (2009) Macro factors in bond risk premia. Rev Financ Stud 22:5027–5067
56. Lütkepohl H, Krätzig M (2004) Applied time series econometrics. Cambridge University Press, Cambridge
57. Luukkonen R, Saikkonen P, Teräsvirta T (1988) Testing linearity against smooth transition autoregressive models. Biometrika 75:491–499
58. Nelson DB (1991) Conditional heteroskedasticity in asset returns: a new approach. Econometrica 59:347–370
59. Paolella MS, Taschini L (2008) An econometric analysis of emission allowance prices. J Bank Finance 32:2022–2032
60. Pippenger MK, Goering GE (1993) A note on the empirical power of unit root tests under threshold processes. Oxf Bull Econ Stat 55:473–481
61. Pippenger MK, Goering GE (2000) Additional results on the power of unit root and cointegration tests under threshold processes. Appl Econ Lett 7:641–644
62. Ploberger W, Kramer W (1992) The CUSUM test with OLS residuals. Econometrica 60:271–285
63. Sadorsky P (2002) Time-varying risk premiums in petroleum futures prices. Energy Econ 24:539–556
64. Sadorsky P (2006) Modeling and forecasting petroleum futures volatility. Energy Econ 28:467–488
65. Stock J, Watson M (2002) Forecasting using principal components from a large number of predictors. J Am Stat Assoc 97:1167–1179
66. Stock J, Watson M (2002) Macroeconomic forecasting using diffusion indexes. J Bus Econ Stat 20:147–162
67. Stock J, Watson M (2005) Implications of dynamic factor models for VAR analysis. NBER working paper 11467, USA
68. Tang K, Xiong W (2009) Index investing and the financialization of commodities. Working Paper, Princeton University, Princeton
69. Teräsvirta T (1994) Specification, estimation, and evaluation of smooth transition autoregressive models. J Am Stat Assoc 89:208–218
70. Teräsvirta T (1998) Modeling economic relationships with smooth transition regressions. In: Ullah A, Giles D (eds) Handbook of applied economic statistics. Dekker, New York, pp 229–246
71. Teräsvirta T (2004) Smooth transition regression modelling. In: Lütkepohl H, Krätzig M (eds) Applied time series econometrics. Cambridge University Press, Cambridge
72. Teräsvirta T, Anderson HM (1992) Characterizing nonlinearities in business cycles using smooth transition autoregressive models. J Appl Econom 7:S119–S136

73. Tong H (1978) On a threshold model. In: Chen CH (ed) Pattern recognition and signal processing. Sijthoff and Noordhoff, Amsterdam, pp 101–141
74. Tong H (1983) Threshold models in non-linear time series analysis. Springer, New York
75. Tong H (1990) Non-linear time series: a dynamical system approach. Clarendon Press, Oxford
76. Tong H, Lim KS (1980) Threshold autoregression, limit cycles and cyclical data (with discussion). J R Stat Soc B 42:245–292
77. Tsay RS (1986) Nonlinearity test for time series. Biometrika 73:461–466
78. Tsay RS (2010) Analysis of financial time series, 3rd edn. Wiley, New York
79. Van Dijk D, Teräsvirta T, Franses PH (2002) Smooth transition autoregressive models—a survey of recent developments. Econom Rev 21:1–47
80. Zakoian JM (1994) Threshold heteroskedastic models. J Econ Dyn Control 18:931–955
81. Zagaglia P (2010) Macroeconomic factors and oil futures prices: a data-rich model. Energy Econ 32:409–417

Chapter 4
The Clean Development Mechanism

Abstract This chapter is dedicated to the Clean Development Mechanism (CDM), and more particularly to the secondary Certified Emissions Reductions (CER) originated under these project mechanisms. Indeed, secondary CERs are valid for compliance within the European trading system up to 13.4% on average. Following a review of the main characteristics of CER contracts and price development, we study the relationship between EUAs and CERs in vector autoregressive and cointegration models. Then, we identify the main CER price drivers based on the Zivot-Andrews structural break tests and regression analysis. Finally, we discuss the main reasons behind the existence of the CER-EUA spread, and highlight the possibilities to benefit from arbitrage opportunities. The Appendix shows how to represent the interactions between EUAs and CERs in a Markov regime-switching environment.

4.1 CERs Contracts and Price Development

According to the article 12 of the Kyoto Protocol, projects under the Clean Development Mechanism (CDM) consist in achieving greenhouse gases emissions reduction in non-Annex B countries. After validation, the CDM Executive Board (CDM EB) of the UNFCCC delivers credits that may be used by Annex B countries for use towards their compliance position. Certified Emissions Reductions (CERs) from CDM projects are credits flowing into the global compliance market generated through emission reductions.

Therefore, the Kyoto Protocol's CDM provides to utilities regulated by the EU ETS the possibility to cut the costs imposed on them by buying relatively cheaper carbon offsets from developing countries, funding emissions cuts there instead. CER prices are determined on the supply side by the decisions of the CDM EB, which decides on the delivery rules. On the demand-side, CER prices are determined by various factors such as the decisions of the European Commission (EC) which determine the institutional fungibility within the European system, and the CERs demand from governments to meet their compliance within the Kyoto Protocol (such as Japan) which absorbs part of the CER demand away from compliance within the EU System.

Albeit being determined on distinct emissions markets, CERs and EUAs may be exchanged based on their representative trading unit. One CER is equal to one ton of CO_2-equivalent emissions reduction, while one EUA is equal to one ton of CO_2

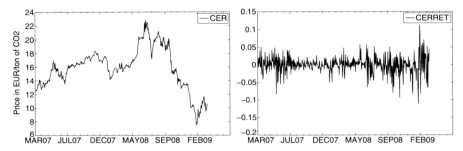

Fig. 4.1 CER: Raw data (*left panel*) and logreturns (*right panel*) from March 09, 2007 to March 31, 2009

Table 4.1 Descriptive statistics for Certified Emissions Reductions in raw (CER) and logreturns (CERRET) transformation

	CER	CERRET
Mean	15.8579	−0.0003
Median	16.0687	0.0001
Maximum	22.8500	0.1125
Minimum	7.4846	−0.1104
Standard Deviation	2.9864	0.0244
Skewness	−0.3514	−0.3703
Kurtosis	3.1352	5.9619
Jarque-Bera (JB)	11.2960	205.0775
Probability (JB)	0.0035	0.0001
Observations	529	528

Source: ECX

emitted in the atmosphere. As soon as they have been sold on exchanges, secondary CERs and EUAs are considered as valid tradable units under the EU ETS. While primary CERs are purchases direct from projects, secondary CERs are free of project risks and traded on exchanges.[1]

Figure 4.1 shows the secondary CER price series exchanged on ECX from March 09, 2007 to March 31, 2009. Prior to the opening of CERs trading on ECX, Reuters provided a similar CER price index. We notice that CERs vary in the range of 15 Euro per ton of CO_2. As for other chapters, the dataset used in this section can be downloaded at: http://sites.google.com/site/jpchevallier/publications/books/springer/data_chapter4.csv.

Besides, we can infer from Table 4.1 that CERs traded as high as 22.85 Euro and as low as 7.48 Euro during our study period. As discussed later in this chapter (see the left panel of Fig. 4.13), the corresponding volume of CERs exchanged during

---

[1]CERs can be traded at different prices depending on the status of the project and the risks involved. It appears important to distinguish forward vs. issued CERs. CERs that are sold in a forward transaction, i.e. before the underlying GHG emissions reductions have been achieved, carry a risk of non-delivery. Thus, they should be exchanged by bilateral trade at a lower price than already issued CERs exchanged on market places, and which correspond to verified emissions reductions.

the period is equal to 584 million ton of CO_2. The price path of CERs therefore seems lower to that of EUAs during the same period. Thus, one of the questions raised in this chapter is the following: why do CERs command a lower price than EU Allowances? We will also discuss the CER price drivers, with the introduction of market-specific variables.

A large inflow of CERs could drive EUA prices down, as it increases allowances supply. Besides, such an inflow poses a threat to EU carbon prices, by undermining the impact of the scheme on driving low-carbon investments in Europe. Thus, an absolute maximum percentage of the effort that will be achievable through the CDM (coupled with quality criteria) appears necessary to maintain the edge of the EU scheme in driving domestic emissions curbs.

The EU Linking Directive allows the import of CERs for compliance into the EU ETS up to 13.4% on average. The import limit is indeed equal to 1.7 billion tons of offsets being allowed into the EU ETS from 2008–2020, i.e. an absolute maximum of 50% of the effort will be achievable through the CDM, coupled with quality and technological criteria. This import limit is also justified by the fact that, despite regulatory uncertainty, it is conceivable that a significant share of the CER import allowances can be banked forward to Phase III of the EU ETS. For instance, in 2008, utilities could not use more than 200 million tons of CERs for compliance.

4.2   Relationship with EU Emissions Allowances

Since the EU ETS represents the world's largest emissions trading system (in terms of market activity and liquidity), it is possible that EUAs have a statistical influence on CER prices. This is also meaningful in an economic context to find that EUAs and sCERs are inter-related, since they both represent the same emissions asset that can be used for arbitrage purposes for compliance within the European trading system. Indeed, the rationale behind the influence of EUAs on CERs channels through compliance mechanisms: while States directly manage their own compliance within the framework of the Kyoto Protocol, secondary CERs can be used by firms for compliance within the EU ETS (up to 13% on average). Therefore, the price path between the two assets—which equally represent one ton of CO_2 emitted in the atmosphere—is expected to react to common drivers.

In this section, we study first the relationship between CERs and EUAs in a vector autoregressive (VAR) framework [20]. Second, we further explore this relationship in the context of linear cointegration [11].

4.2.1 VAR Analysis

Let $Z_t = \begin{pmatrix} X_t^1 \\ X_t^2 \end{pmatrix}$ be the vector process formed of the two (properly transformed to stationary) EUA and sCER prices. Then, the VAR(p) model reads:

$$Z_t = C + \Gamma_1 Z_{t-1} + \cdots + \Gamma_p Z_{t-p} + \varepsilon_t \tag{4.1}$$

Table 4.2 R Code: VAR EUA-CER

```
1 path<-"C:/"
2 setwd(path)
3 library(vars)
4 data=read.csv("data_chapter4.csv",sep=",")
5 attach(data)
6 eua=data[,2]
7 cer=data[,3]
8 euacer<-cbind(eua,cer)
9 VARselect(euacer, lag.max=8, type="const")
10 vareuacer <- VAR(euacer, p=1, type="const")
11 summary(vareuacer, equation = "eua")
12 summary(vareuacer, equation = "cer")
13 plot(vareuacer, names ="eua")
14 plot(vareuacer, names ="cer")
15 portmanteau <- serial.test(vareuacer, lags.pt=16,type="PT.asymptotic")
16 normality <- normality.test(vareuacer)
17 arch <- arch.test(vareuacer, lags.multi=5)
18 plot(stability(vareuacer), nc=2)
19 causality(vareuacer, cause = c("eua"))
20 causality(vareuacer, cause = c("cer"))
21 irf.vareuacer1 <- irf(vareuacer, impulse = "cer", response = "eua",
n.ahead=10, ortho=FALSE, cumulative=FALSE, boot=FALSE)
22 plot(irf.vareuacer1)
23 irf.vareuacer2 <- irf(vareuacer, impulse = "eua", response = "cer",
n.ahead=10, ortho=FALSE, cumulative=FALSE, boot=FALSE)
24 plot(irf.vareuacer2)
25 fevd.vareuacer <- fevd(vareuacer, n.ahead=10)
26 plot(fevd.vareuacer,addbars=2)
```

where $C = \binom{C^1}{C^2}$ is a constant vector, $\Gamma_1, \ldots, \Gamma_p$ are 2×2 matrices and the vector process $\varepsilon_t = \binom{\varepsilon_t^1}{\varepsilon_t^2}$ is formed of independent random variables following a centered bi-variate normal distribution $N(0, \sum)$.

The calibration of the model VAR(p) proceeds in three classic steps:

1. the optimal order p is selected using an information criterion, which is an indicator of the relevance of a model, giving a positive weight to the likelihood of the model and a negative weight to the number of model parameters; e.g., the Schwarz Information Criterion (SIC) is equal to $2LL - \ln(T)n$, where LL is the log-likelihood, T is the number of observations and n is the number of parameters of the model.
2. once p is known, $C, \Gamma_1, \ldots, \Gamma_p$ are determined by OLS.
3. lastly, the standard deviation and correlation of the residuals give the matrix $\sum$.

We perform these various steps based on the R code shown in Table 4.2.
Lines 1 to 8 allow to select the relevant data.

4.2.1.1 Selecting the Order of the VAR(p)

Line 9 allows to choose the order p of the VAR based on information criteria. The output is reproduced in Table 4.3.

Based on the Schwarz criterion, we choose to implement the most parsimonious VAR(p) model of order 1 (VAR(1)).

Table 4.3   R Output: Selection of the order of the VAR(p)

```
$selection
AIC(n) HQ(n) SC(n) FPE(n)
 6   1   1   6

$criteria
       1           2           3           4           5
AIC(n) -1.467096e+01 -1.466889e+01 -1.467621e+01 -1.468344e+01 -1.469126e+01
HQ(n)  -1.465177e+01 -1.463689e+01 -1.463142e+01 -1.462585e+01 -1.462087e+01
SC(n)  -1.462195e+01 -1.458720e+01 -1.456185e+01 -1.453641e+01 -1.451155e+01
FPE(n)  4.250904e-07  4.259749e-07  4.228673e-07  4.198209e-07  4.165557e-07
       6           7           8
AIC(n) -1.469439e+01 -1.468394e+01 -1.467186e+01
HQ(n)  -1.461120e+01 -1.458795e+01 -1.456307e+01
SC(n)  -1.448201e+01 -1.443889e+01 -1.439413e+01
FPE(n)  4.152570e-07  4.196238e-07  4.247314e-07
```

4.2.1.2 VAR(1) Estimation Results

Lines 10 to 12 lead to the vector autoregression estimates reproduced in Table 4.4.

We observe that lagged sCER prices impact EUA prices (at the 5% level), and that sCER prices are best explained by an AR(1) at the 10% significance level. Therefore, we find that EUAs do not have a statistically significant impact on CERs in this example.

4.2.1.3 Diagnostic Tests

To confirm that the VAR(1) model for EUAs and CERs is well specified, Lines 13 and 14 conduct residual analysis.

From a visual inspection of Figs. 4.2 and 4.3, the residuals do not appear to be autocorrelated. Indeed, we cannot detect any pattern in the ACF and PACF functions, which stay safely within the bounds of the confidence intervals.

Box 4.1 (Portmanteau Test of a VAR(p))   For testing the lack of serial correlation in the residuals of a VAR(p), the Portmanteau statistic is defined as:

$$Q_h = T \sum_{j=1}^{h} \mathrm{tr}(\hat{C}'_j \hat{C}_0^{-1} \hat{C}_j \hat{C}_0^{-1}) \qquad (4.2)$$

with Q_h the Breusch-Godfrey Q-statistic at lag h, T the number of observations, tr the trace, and $\hat{C}_i = \frac{1}{T} \sum_{t=i+1}^{T} \hat{\mathbf{u}}_t \hat{\mathbf{u}}'_{t-i}$. $\mathbf{u}_t$ is a K-dimensional white noise process with time invariant positive definite covariance matrix $E(\mathbf{u}_t \mathbf{u}'_t) = \sum \mathbf{u}$. The test statistic has an approximate $\chi^2(K^2 h - n^*)$ distribution, where n^* is the number of coefficients excluding deterministic terms of the VAR(p).

Table 4.4 R Output: VAR EUA-CER Estimates

```
VAR Estimation Results:
=========================
Endogenous variables: eua, cer
Deterministic variables: const
Sample size: 528
Log Likelihood: 2385.512
Roots of the characteristic polynomial:
0.1007 0.1007
Call:
VAR(y = euacer, p = 1, type = "const")

Estimation results for equation eua:
====================================
eua = eua.l1 + cer.l1 + const

Estimate Std. Error t value Pr(>|t|)
eua.l1 -0.0286791 0.1108578 -0.259 0.7960
cer.l1 0.2805998 0.1217078 2.306 0.0215 *
const 0.0006586 0.0015567 0.423 0.6724
---
Signif. codes: 0 '***' 0.001 '**' 0.01 '*' 0.05 '.' 0.1 ' ' 1

Residual standard error: 0.03576 on 525 degrees of freedom
Multiple R-Squared: 0.02943, Adjusted R-squared: 0.02573
F-statistic: 7.959 on 2 and 525 DF, p-value: 0.0003935

Estimation results for equation cer:
====================================
cer = eua.l1 + cer.l1 + const

Estimate Std. Error t value Pr(>|t|)
eua.l1 -0.0524008 0.0755892 -0.693 0.4885
cer.l1 0.1594293 0.0829873 1.921 0.0553 .
const -0.0002376 0.0010614 -0.224 0.8230
---
Signif. codes: 0 '***' 0.001 '**' 0.01 '*' 0.05 '.' 0.1 ' ' 1

Residual standard error: 0.02439 on 525 degrees of freedom
Multiple R-Squared: 0.01304, Adjusted R-squared: 0.009277
F-statistic: 3.467 on 2 and 525 DF, p-value: 0.03191
```

Lines 15 to 17 are able to confirm statistically this first diagnostic test, as included in Table 4.5.

Portmanteau refers to the asymptotic Portmanteau test with a maximum lag of 16, *ARCH* is the multivariate ARCH test with a maximum lag of order 5, *JB* is the Jarque-Bera normality test for multivariate series applied to the residuals of the VAR(1). Kurtosis and skewness stand for separate tests for multivariate skewness and kurtosis. df stands for degree of freedom of the test statistic.

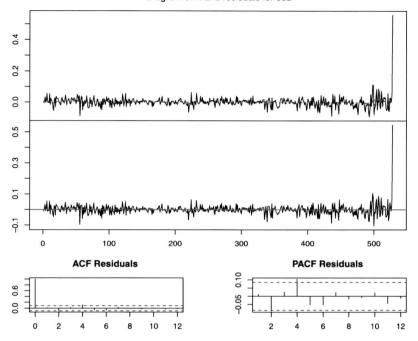

Fig. 4.2 VAR EUA-CER: Diagram of fit (*upper panels*) and residuals (*lower panels*) for the EUA Equation

Box 4.2 (Multivariate Jarque-Bera Statistic) The Jarque-Bera statistic for the multivariate case is defined as:

$$JB_{mv} = s_3^2 + s_4^2 \tag{4.3}$$

s_3^2 and s_4^2 are computed as:

$$s_3^2 = T\mathbf{b}_1'\mathbf{b}_1/6 \tag{4.4}$$

$$s_4^2 = T(\mathbf{b}_2 - 3_K)'(\mathbf{b}_2 - 3_K)/24 \tag{4.5}$$

with $\mathbf{b}_1$ and $\mathbf{b}_2$ the third and fourth non-central moment vectors of the standardized residuals $\hat{\mathbf{u}}_t^s = \tilde{P}^-(\hat{\mathbf{u}}_t - \bar{\hat{\mathbf{u}}}_t)$. The Cholesky decomposition of the residual covariance matrix $\tilde{P}$ is a lower triangular matrix with positive diagonal such that $\tilde{P}\tilde{P}' = \sum \mathbf{u}$. The test statistic JB_{mv} is distributed as $\chi^2(2K)$. Note the multivariate skewness (s_3^2) and kurtosis (s_4^2) tests are distributed as $\chi^2(K)$.

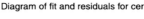

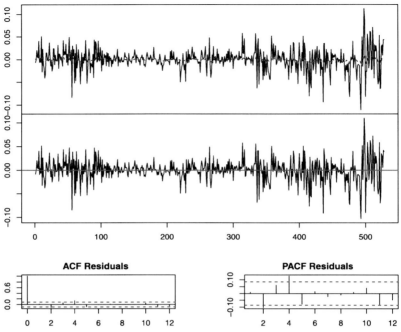

Fig. 4.3 VAR EUA-CER: Diagram of fit (*upper panels*) and residuals (*lower panels*) for the CER Equation

We verify that the residuals are not autocorrelated (as the *p*-value of the Portmanteau test is strictly above 5%). Furthermore, in our example, the null hypothesis of normality can be rejected (at the 1% level). Similarly, the null hypothesis of homoscedasticity can be rejected (at the 1% level), indicating ARCH effects.[2]

Box 4.3 (Multivariate ARCH Test)   The multivariate ARCH test is based on the following regression [6, 7, 13]:

$$\mathrm{vech}(\hat{\mathbf{u}}_t\hat{\mathbf{u}}_t') = \beta_0 + B_1\,\mathrm{vech}(\hat{\mathbf{u}}_{t-1}\hat{\mathbf{u}}_{t-1}') + \cdots + B_q\,\mathrm{vech}(\hat{\mathbf{u}}_{t-q}\hat{\mathbf{u}}_{t-q}') + \mathbf{v}_t \quad (4.6)$$

[2]See [15] for an econometric analysis of EUAs and CERs including GARCH effects.

Table 4.5   R Output: Diagnostic tests for the residuals of the VAR EUA-CER

```
Portmanteau Test (asymptotic)

data: Residuals of VAR object vareuacer
Chi-squared = 76.7209, df = 60, p-value = 0.07165

JB-Test (multivariate)

data: Residuals of VAR object vareuacer
Chi-squared = 273835.2, df = 4, p-value < 2.2e-16

Skewness only (multivariate)

data: Residuals of VAR object vareuacer
Chi-squared = 4952.619, df = 2, p-value < 2.2e-16

Kurtosis only (multivariate)

data: Residuals of VAR object vareuacer
Chi-squared = 268882.6, df = 2, p-value < 2.2e-16

ARCH (multivariate)

data: Residuals of VAR object vareuacer
Chi-squared = 163.0051, df = 45, p-value = 2.776e-15
```

with v_t a spherical error process, and vech is the column-stacking operator for symmetric matrices that stacks the columns from the main diagonal on downwards. The dimension of β_0 is $\frac{1}{2}K(K+1)$ and for the coefficient matrices B_i with $i = 1, \ldots, q$, $\frac{1}{2}K(K+1) \times \frac{1}{2}K(K+1)$. The null hypothesis is $H_0 : B_1 = B_2 = \cdots = B_q = 0$. The test statistic is defined as:

$$VARCH_{LM}(q) = \frac{1}{2}TK(K+1)R_m^2 \qquad (4.7)$$

with:

$$R_m^2 = 1 - \frac{2}{K(K+1)} \operatorname{tr}(\hat{\Omega}\hat{\Omega}_0^{-1}) \qquad (4.8)$$

$\hat{\Omega}$ denotes the covariance matrix. The test statistic is distributed as $\chi^2(qK^2(K+1)^2/4)$.

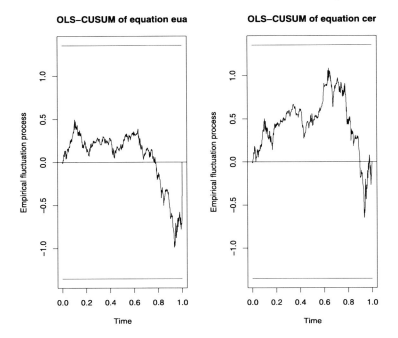

Fig. 4.4 VAR EUA-CER: OLS-CUSUM Test for EUA (*left panel*) and CER (*right panel*)

4.2.1.4 Stability

> **Box 4.4** (OLS-Based CUSUM Processes)   To check for structural changes in
> a time series model, we can use the OLS-based CUSUM process [19] which
> contains cumulative sums of standardized residuals:
>
> $$W_n^0(t) = \frac{1}{\hat{\sigma}\sqrt{n}} \sum_{i=1}^{\lfloor nt \rfloor} \hat{u}_i \quad (0 \leq t \leq 1) \qquad (4.9)$$
>
> with $W_n^0(t)$ the limiting process for the standard Brownian bridge $W^0(t) = w(t) - tW(1)$, and $W(.)$ the standard Brownian motion. Under a single-shift
> alternative, the process should have a peak around the breakpoint.

Line 18 checks that the VAR estimation results obtained are not contaminated by
structural changes, by resorting to the OLS CUSUM test [19].

Figure 4.4 gives the various processes together with their boundaries at an
(asymptotic) 5% significance level for the EUA and CER equations. While we can
observe some peaks for each time series during the period under consideration,

Table 4.6 R Output: Granger Causality Tests for the VAR EUA-CER

```
Granger causality H0: eua do not Granger-cause cer

data: VAR object vareuacer
F-Test = 0.4806, df1 = 1, df2 = 1050, p-value = 0.4883

Granger causality H0: cer do not Granger-cause eua

data: VAR object vareuacer
F-Test = 5.3154, df1 = 1, df2 = 1050, p-value = 0.02133
```

none of them indicate a clear structural shift. Visually, the OLS-based CUSUM processes remain safely within their boundaries, which further confirms that our VAR(1) model is correctly specified.

4.2.1.5 Causality in the Granger Sense

Lines 19–20 conduct Granger causality tests between EUAs and CERs, which yields to the output included in Table 4.6.

Recall that a process P_t^1 Granger causes P_t^2 at the order p if, in the linear regression of P_t^2 on lagged prices $P_{t-1}^1, \ldots, P_{t-p}^1, P_{t-1}^2, \ldots, P_{t-p}^2$, at least one of the regression coefficients of P_t^1 on the lagged prices $P_{t-1}^2, \ldots, P_{t-p}^2$ is significantly different from 0.

Granger causality is examined using the Granger causality test testing the null hypothesis H_0 that all regression coefficients of P_t^1 on the lagged prices $P_{t-1}^2, \ldots, P_{t-p}^2$ are null. A p-value lower than 0.05 means that H_0 can be rejected (and causality accepted) with 95% confidence level.

In our practical example, we identify that a positive causality runs from CERs to EUAs, but not conversely.[3] Hence, the Granger causality test confirms the previous analysis that—in this example—CERs have a statistically significant impact on EUAs.

Since Granger causality test results are also useful to determine the order of the Cholesky decomposition in the impulse response function analysis, it thus follows very logically to simulate random shocks in the next section in order to have a better understanding of the interrelationships between the two variables.

---

[3] As usual, the user may check whether the results obtained are sensitive to the order of the lag retained for the Granger causality test.

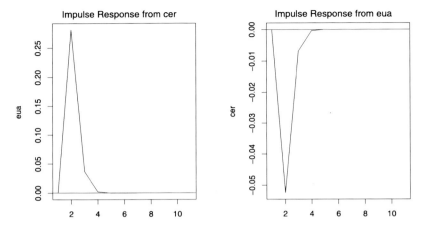

Fig. 4.5 VAR EUA-CER: Impulse response functions for CER → EUA (*left panel*) and EUA → CER (*right panel*)

4.2.1.6 Impulse Response Function Analysis

Lines 21–24 allow to perform an impulse response function (IRF) analysis between EUAs and CERs. As is standard, the orthogonalized impulse responses are derived from a Cholesky decomposition of the error variance-covariance matrix $\sum \mathbf{u} = PP'$, with P being a lower triangular matrix [7]. This Cholesky decomposition is drawn from the Granger causality tests, with EUA being defined as the most exogenous variable in our system. Note that a different ordering of the variables might produce different outcomes with respect to the impulse responses.

Figure 4.5 shows the results of the IRF analysis. We observe that EUAs react rapidly (at time $t = 2$) and positively to a shock on CERs in the system. The effects of the initial shock are dampened are time $t = 3$, and disappear at time $t = 4$. Besides, we may note that CERs tend to be negatively affected (with a lag of two periods) by a shock on EUAs, and that the effect of this shock disappears after four periods. The magnitude of the effect is much stronger for EUAs (up to 0.25 standard deviation) than for CERs (-0.05 standard deviation). In this example, both variables are therefore found to exhibit an asymmetric pattern with respect to each other in terms of responses to exogenous shocks. Note that the oscillation towards zero is characteristic of stationary VAR models.

Finally, Lines 25–26 perform variance decomposition analysis for the pair of EUAs and CERs. In Fig. 4.6, we observe that the variance of the forecast error of the EUA variable is due to 99% to its own innovations, and to 1% to the innovations of the CER variable. The variance of the forecast error of the sCER variable is due to 45% to the EUA variable, and to 55% to itself. This graph therefore provides another view concerning the statistical influences running between EUA and CER prices.

We extend our analysis of the relationship between EUAs and CERs in the next section by introducing the concept of cointegration.

Fig. 4.6 VAR EUA-CER:
Forecast Error Variance
Decomposition for EUA
(*upper panel*) and CER
(*lower panel*)

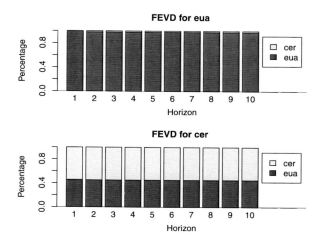

4.2.2 Cointegration

In what follows, we wish to use the concept of cointegration to look for a stationary linear combination of EUA and sCER prices, which will represent the long-run equilibrium. In this stylized exercise, we will then study the error-correction mechanisms insuring the reversion to the long-run equilibrium.

4.2.2.1 Preliminary Conditions

As a pre-requisite condition for cointegration, the time series need to be integrated of the same order. The reader can check, based on standard stationarity tests, that the prices of EUA and sCER are non stationary and integrated of order one ($I(1)$). This amounts to checking that they are difference stationary, i.e. ΔX_t^e and $\Delta X_t^{e'}$ are stationary.

4.2.2.2 Johansen Cointegration Tests

As is standard in a linear cointegration exercise, the econometrican needs to check first if the EUA and sCER variables are cointegrated, i.e. if β exists such that $R_t = X_t^e - \beta X_t^{e'}$ is stationary. This can be done by performing an OLS regression of X_t^e on $X_t^{e'}$, or more rigorously by using the Johansen cointegration test [9, 10].

Let X_t be a vector of N variables, all $I(1)$:

$$X_t = \Phi_1 X_{t-1} + \cdots + \Phi_p X_{t-p} + \varepsilon_t \tag{4.10}$$

with $\varepsilon_t \sim WGN(0, \Omega)$, WGN denotes the White Gaussian Noise, Ω denotes the variance covariance matrix, and Φ_i ($i = 1, \ldots, p$) are parameter matrices of size ($N \times N$).

Table 4.7 R Code: Johansen Cointegration Tests

```
1 path<-"C:/"
2 setwd(path)
3 library(vars)
4 data=read.csv("data.csv",sep=",")
5 attach(data)
6 logeua=data[,2]
7 logcer=data[,3]
8 price.mat <- data.frame(logeua,logcer)
9 vecm.trace <- ca.jo(price.mat, type='trace')
10 vecm.eigen <- ca.jo(price.mat, type='eigen')
```

Under the null H_0, there exists r cointegration relationships between N variables, i.e. X_t is cointegrated with rank r.

The Johansen cointegration tests can be performed on the logarithmic transformation of EUAs and CERs in R,[4] as shown in Table 4.7.

Lines 1 to 8 load the relevant data in matrix form. Lines 9 and 10 compute the trace and maximum eigenvalue tests, respectively.

4.2.2.3 Error-Correction Model

The next step of the cointegration model consists in describing the dynamics of EUAs and sCERs in terms of the residuals of the long-term relation [8]. We want to introduce an error-correction mechanism on the levels and on the slopes between the variables e and e':

$$\begin{pmatrix} \Delta X_t^e \\ \Delta X_t^{e'} \end{pmatrix} = \begin{pmatrix} \mu_e \\ \mu_{e'} \end{pmatrix} + \sum_{k=1}^{p} \Gamma_k \begin{pmatrix} \Delta X_{t-k}^e \\ \Delta X_{t-k}^{e'} \end{pmatrix} + \begin{pmatrix} \Pi_e \\ \Pi_{e'} \end{pmatrix} R_t + \begin{pmatrix} \varepsilon_t^e \\ \varepsilon_t^{e'} \end{pmatrix} \tag{4.11}$$

where

- e stands for EUA, and e' stands for sCER;
- X_t^e is the log price of variable e at time t;
- the 2×1 vector process $\Delta Z_t = (\Delta X_t^e = X_{t+1}^e - X_t^e, \Delta X_t^{e'} = X_{t+1}^{e'} - X_t^{e'})'$ is the vector of EUA and sCER price returns;
- $\mu = (\mu_{X,e}, \mu_{X,e'})$ is the 1×2 vector composed of the constant part of the drifts;
- Γ_k are 2×2 matrices expressing dependence on lagged returns;
- $(R_t = X_t^e - \beta X_t^{e'})$ is the process composed of the deviations to the long-term relation between the EUA and sCER log prices;
- Π is a 2×1 vector matrix expressing the sensitivity to the deviations to the long-term relation between the EUA and sCER prices;

---

[4]Note the time series of log EUAs and log CERs are not provided in the data for this chapter.

- the residual shocks $(\varepsilon_t^e, \varepsilon_t^{e'})$ are assumed to be i.i.d. with a centered bi-variate normal distribution $N(0, \sum)$.

However, by considering a purely linear model, it is possible that the econometrician will either misspecify the model, or ignore a valid a valid cointegration relationship. That is why we detail below the cointegration methodology with an unknown structural break.

4.2.2.4   VECM and Structural Shift

In this section, we explore the possibility of wrongly accepting a cointegration relationship, when some of the underlying time series are contaminated by a structural break. We present the procedure for estimating a vector error-correction model (VECM) with a structural shift in the level of the process, as developed by Lütkepohl, Saikkonen and Trenkler (2004, [12]). By doing so, we draw on the notations by Pfaff (2008, [17]).

Let $\mathbf{y_t}$ be a $K \times 1$ vector process generated by a constant, a linear trend, and level shift terms:[5]

$$\mathbf{y_t} = \boldsymbol{\mu_0} + \boldsymbol{\mu_1} t + \delta d_{t\tau} + \mathbf{x_t} \qquad (4.12)$$

with $d_{t\tau}$ a dummy variable which takes the value of one when $t \geq \tau$, and zero otherwise. The shift point τ is unknown, and is expressed as a fixed fraction of the sample size:

$$\tau = [T\lambda], \quad 0 < \underline{\lambda} \leq \lambda \leq \overline{\lambda} < 1 \qquad (4.13)$$

where $\underline{\lambda}$ and $\overline{\lambda}$ define real numbers, and $[\cdot]$ the integer part. Therefore, the shift cannot occur at the very beginning or the very end of the sample. The estimation of the structural shift is based on the regressions:

$$\mathbf{y_t} = \mathbf{v_0} + \mathbf{v_1} t + \delta d_{t\tau} + \mathbf{A_1} \mathbf{y_{t-1}} + \cdots + \mathbf{A_1} \mathbf{y_{t-p}} + \varepsilon_{t\tau}, \quad t = p+1, \dots, T \quad (4.14)$$

with $\mathbf{A_i}, i = 1, \dots, p$ the $K \times K$ coefficient matrices, and ε_t the spherical K-dimensional error process. The estimator for the breakpoint is defined as:

$$\hat{\tau} = \arg \min_{\tau \in \mathcal{T}} \det \left(\sum_{t=p+1}^{T} \hat{\boldsymbol{\varepsilon}}_{\mathbf{t\tau}} \hat{\boldsymbol{\varepsilon}}_{\mathbf{t\tau}}' \right) \qquad (4.15)$$

with $\mathcal{T} = [T\underline{\lambda}, T\overline{\lambda}]$, and $\hat{\boldsymbol{\varepsilon}}_{\mathbf{t\tau}}$ the least squares residuals of Eq. (4.14). Once the breakpoint $\hat{\tau}$ has been estimated, the data are adjusted as follows:

$$\hat{\mathbf{x}}_{\mathbf{t}} = \mathbf{y_t} - \hat{\boldsymbol{\mu}}_{\mathbf{0}} - \hat{\boldsymbol{\mu}}_{\mathbf{1}} t + \hat{\delta} d_{t\hat{\tau}} \qquad (4.16)$$

---

[5]Note that [12] develop their analysis in the context where $\mathbf{x_t}$ can be represented as a VAR(p), whose components are at most $I(1)$ and cointegrated with rank r.

Table 4.8 R Code: Johansen Cointegration Test with Structural Shift

```
1 path<-"C:/"
2 setwd(path)
3 library(vars)
4 data=read.csv("data.csv",sep=",")
5 attach(data)
6 logeua=data[,2]
7 logcer=data[,3]
8 price.mat <- data.frame(logeua,logcer)
9 cointegration.break <- summary(cajolst (price.mat))
```

The test statistic writes:

$$LR(r) = T \sum_{j=r+1}^{N} \ln(1 + \hat{\lambda}_j) \qquad (4.17)$$

with corresponding critical values found in [21]. Table 4.8 shows how to perform this task in R, based on the logarithmic transformation of EUAs and CERs prices.[6]

The output returns the trace test statistics, as shown in Table 4.9.

According to these results, we accept the presence of at least one cointegrating relationship ($r = 1$) between EUAs and CERs—considered in logarithmic form—when taking explicitly into account a structural shift in the level of the process (see [15] for another example).

4.2.2.5 Estimation of the VECM

The error-correction model (ECM) writes:

$$\Delta X_t = \Pi_1 \Delta X_{t-1} + \cdots + \Pi_{p-1} \Delta X_{t-p+1} + \Pi_p X_{t-p} + \varepsilon_t \qquad (4.18)$$

where the matrices Π_i ($i = 1, \ldots, p$) are of size ($N \times N$). All variables are $I(0)$, except X_{t-p} which is $I(1)$. For all variables to be $I(0)$, $\Pi_p X_{t-p}$ needs to be $I(0)$ as well.

Let $\Pi_p = -\beta \alpha'$, where α' is an (r, N) matrix which contains r cointegration vectors, and β is an (N, r) matrix which contains the weights associated with each vector. If there exists r cointegration relationships, then $Rk(\Pi_p) = r$. Johansen's cointegration tests are based on this condition. We can thus rewrite Eq. (4.18):

$$\Delta X_t = \Pi_1 \Delta X_{t-1} + \cdots + \Pi_{p-1} \Delta X_{t-p+1} - \beta \alpha' X_{t-p} + \varepsilon_t \qquad (4.19)$$

The VECM can be estimated in R with the code given in Table 4.10.

---

[6]Recall that these price series are not included in the dataset of this chapter.

Table 4.9   R Output: Johansen Cointegration Test with Structural Shift

```
######################
# Johansen-Procedure #
######################

Test type: trace statistic, with linear trend in shift correction

Eigenvalues (lambda):
[1] 0.02161897 0.01012411

Values of teststatistic and critical values of test:

test 10pct 5pct 1pct
r <= 1 | 5.31 5.42 6.79 10.04
r = 0 | 16.58 13.78 15.83 19.85

Eigenvectors, normalized to first column:
(These are the cointegration relations)

logeua.l1 logcer.l1
logeua.l1 1.0000000 1.000000
logcer.l1 -0.4901037 -1.515450

Weights W:
(This is the loading matrix)

logeua.l1 logcer.l1
logeua -0.05968590 0.007171102
logcer -0.04361258 0.018227392
```

Lines 1 to 8 load the data of EUA and CER logreturns in matrix form. Line 9 performs the estimation of the VECM through maximum likelihood methods [9, 10]. This operation produces the output shown in Table 4.11.

Table 4.11 reveals the error correction mechanism which leads towards the long-term stationary relationship between EUAs and sCERs. We observe that EUA and sCER returns indeed correct the deviations to the long-term equilibrium. By combining linearly the short-term variations of EUAs and sCERs, the vector error correction mechanism allows by definition to diminish the fluctuation errors in order to achieve the cointegrating relationship between both variables. The error correction coefficient (*ect*1) estimates indicate a slow adjustment of short-term deviations to the long-term relationship. The VECM explains both EUA and CER price series by their own lagged values (one day lag *eua.dl*1 and *cer.dl*1). It is interesting to see that in the long-run EUAs and CERs move together according to the cointegration relationship estimated by a relatively simple dynamic repercussion. Besides, note that the coefficient for the EUA variable (0.039) is roughly equivalent to the coefficient for the CER variable (0.049). Hence, we cannot conclude with certainty in this

Table 4.10 R Code: VECM

```
1 path<-"C:/"
2 setwd(path)
3 library(vars)
4 data=read.csv("data_chapter4.csv",sep=",")
5 attach(data)
6 eua=data[,2]
7 cer=data[,3]
8 price.mat <- data.frame(eua,cer)
9 vecm.r1 <- cajorls(vecm.eigen,r=1)
10 vecm.level <- vec2var(vecm.eigen,r=1)
11 serial.test.vecm.level <- serial.test(vecm.level, lags.pt=16,
type="PT.asymptotic")
12 normality.test(vecm.level)
13 arch.test(vecm.level)
14 plot(serial.test.vecm.level,names="resids of eua")
15 plot(serial.test.vecm.level,names="resids of cer")
```

Table 4.11 R Output: VECM

```
Call:
lm(formula = substitute(form1), data = data.mat)

Coefficients:
eua.d cer.d
ect1 0.0399704 0.0490506
constant 0.0008016 -0.0002413
eua.dl1 -0.3943986 0.0471124
cer.dl1 -0.2784667 -0.9214969

$beta
ect1
eua.l2 1.00000
cer.l2 -20.83647
```

example that either the EUA price or the CER price is the leader in the long-term price discovery.

4.2.2.6 Diagnostic Tests

Lines 11 to 13 compute diagnostic tests for the residuals of the VECM.

As for the VAR analysis, we find in Table 4.12 that the residuals are not normally distributed and the presence of ARCH effects. Lines 14 and 15 provide further tools for residual analysis.

While Figs. 4.7 and 4.8 do not seem to indicate visually that the residuals are autocorrelated, the results in Table 4.12 stand in sharp contrast as the p-value of the Ljung-Box-Pierce Test is equal to 2.86%. The reader is invited to reconsider

Table 4.12 R Output: VECM EUA-CER Diagnostic Tests

```
Portmanteau Test (asymptotic)

data: Residuals of VAR object vecm.level
Chi-squared = 80.1577, df = 58, p-value = 0.02860

JB-Test (multivariate)

data: Residuals of VAR object vecm.level
Chi-squared = 173191.8, df = 4, p-value < 2.2e-16

$Skewness

Skewness only (multivariate)

data: Residuals of VAR object vecm.level
Chi-squared = 3459.678, df = 2, p-value < 2.2e-16

$Kurtosis

Kurtosis only (multivariate)

data: Residuals of VAR object vecm.level
Chi-squared = 169732.1, df = 2, p-value < 2.2e-16

ARCH (multivariate)

data: Residuals of VAR object vecm.level
Chi-squared = 145.8837, df = 45, p-value = 1.403e-12
```

the presence of a cointegration relationship between the two time series with an extended dataset (see also [15] and [3] for further analyses on this topic).

Overall, this section has introduced useful tools (VAR and cointegration) in order to study jointly the time series of EUAs and CERs (see [18] for another application of the VECM approach to carbon, electricity and fuel variables). In addition, note that the DCC MGARCH model could also appear as a suitable econometric framework in order to track the relationship between these two variables (see [4] for an application).

Next, we investigate empirically the main CER price drivers.

4.3 CERs Price Drivers

This section contains (i) an exposition of the Zivot-Andrews structural break test applied to the CER price series, and (ii) a regression analysis to identify the main CER price drivers.

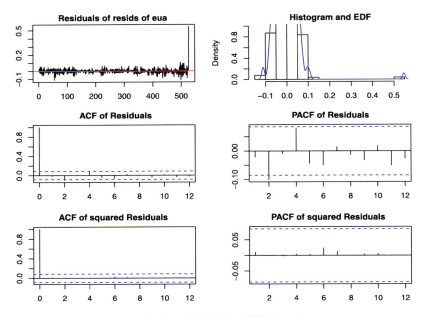

Fig. 4.7   VECM EUA-CER: Residuals analysis for the EUA Equation

4.3.1 Zivot-Andrews Structural Break Test

As for EUAs, we may investigate the main shocks affecting the CER price path
based on structural break tests, and attempt to relate them to actual market developments or institutional decision changes. We introduce below the well-known test by
Zivot and Andrews (1992, [22]). In what follows, we draw notations from [17].

4.3.1.1 Setting of the Test

This test allows to determine endogenously the most likely occurrence of a structural
break. The estimation procedure chooses the structural break date as the point in
time associated with the least favorable result for the null hypothesis of a random
walk with drift. The test statistic builds on Perron's (1989, [16]) Student t ratio:

$$t_{\hat{\alpha}^i}[\hat{\lambda}^i_{\text{inf}}] = \inf_{\lambda \in \Delta} t_{\hat{\alpha}^i}(\lambda), \quad i = A, B, C \tag{4.20}$$

with $1 < T_\tau < T$, $\lambda = \frac{T_\tau}{T}$ the fraction of the structural break point with respect to the
total sample, and Δ a closed subset of $(0, 1)$. Following ADF-type tests, the three
models feature: A a one-time shift in the levels of the series, B a change in the rate
of growth, and C a combination of both. Depending on the model selected, the test

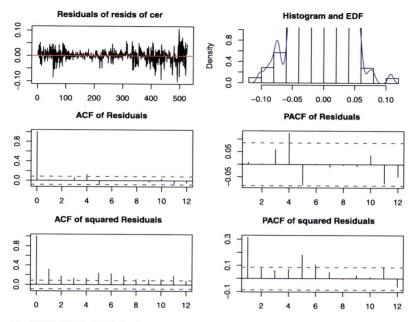

Fig. 4.8   VECM EUA-CER: Residuals analysis for the CER Equation

statistic is computed from one of the following regressions:

$$y_t = \hat{\mu}^A + \hat{\theta}^A DU_t(\hat{\lambda}) + \hat{\beta}^A t + \hat{\alpha}^A y_{t-1} + \sum_{i=1}^{k} \hat{c}_i^A \Delta y_{t-i} + \hat{\varepsilon}_t \qquad (4.21)$$

$$y_t = \hat{\mu}^B + \hat{\beta}^B t + \hat{\gamma}^B DT_t^*(\hat{\lambda}) + \hat{\alpha}^B y_{t-1} + \sum_{i=1}^{k} \hat{c}_i^B \Delta y_{t-i} + \hat{\varepsilon}_t \qquad (4.22)$$

$$y_t = \hat{\mu}^C + \hat{\theta}^C DU_t(\hat{\lambda}) + \hat{\beta}^C t + \hat{\gamma}^C DT_t^*(\hat{\lambda}) + \hat{\alpha}^C y_{t-1} + \sum_{i=1}^{k} \hat{c}_i^C \Delta y_{t-i} + \hat{\varepsilon}_t \quad (4.23)$$

with $DU_t(\lambda) = 1$ if $t > T\lambda$ (and zero otherwise), $DT_t^*(\lambda) = t - T\lambda$ for $t > T\lambda$ (and zero otherwise). Zivot and Andrews (1992, [22]) provide the corresponding critical values.

4.3.1.2 Application to CERs

The R code contained in Table 4.13 allows to perform the Zivot-Andrews structural break test for the CER price series.

Lines 1 to 6 load the data. Line 7 computes the Zivot-Andrews test statistic for model C, as shown in Table 4.14.

Table 4.13 R Code: Zivot-Andrews Structural Break Test for CERs

```
1 path<-"C:/"
2 setwd(path)
3 library(vars)
4 data=read.csv("data_chapter4.csv",sep=",")
5 attach(data)
6 cer=data[,3]
7 za.cer <- ur.za(cer,model="both",lag=8)
8 plot(za.cer)
```

Fig. 4.9 Zivot-Andrews test statistic for the CER price series

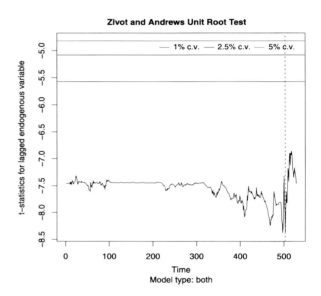

The model appears to be well-estimated, as most parameters of interest in Eq. (4.23) are found to be statistically significant. Besides, the break date indicated corresponds to February 23, 2009. This breakpoint may be due to a delayed effect of the financial crisis on the CER market [15]. Figure 4.9 provides a visual confirmation of these test statistics results.

Next, we explore the CER price drivers.

4.3.2 Regression Analysis

Similarly to EUAs, we may uncover the main CER price drivers based on a regression analysis. As explained at the beginning of this chapter, CERs have essentially the same drivers as EUAs, due to the ability of operators to use CERs for compliance within the EU ETS. Hence, we follow [15] by specifying the following regression

Table 4.14   R Output: Zivot-Andrews Structural Break Test for CERs

```
###############################
# Zivot-Andrews Unit Root Test #
###############################

Call:
lm(formula = testmat)

Residuals:
Min 1Q Median 3Q Max
-0.099543 -0.010312 0.000764 0.012998 0.098546

Coefficients:
Estimate Std. Error t value Pr(>|t|)
(Intercept) 4.710e-03 2.299e-03 2.049 0.04097 *
y.l1 -1.139e-01 1.327e-01 -0.858 0.39128
trend -2.257e-05 8.074e-06 -2.796 0.00538 **
y.dl1 2.223e-01 1.243e-01 1.788 0.07430 .
y.dl2 1.095e-01 1.161e-01 0.944 0.34582
y.dl3 1.574e-01 1.088e-01 1.448 0.14834
y.dl4 2.600e-01 9.952e-02 2.613 0.00924 **
y.dl5 1.398e-01 9.072e-02 1.541 0.12384
y.dl6 1.266e-01 7.764e-02 1.631 0.10348
y.dl7 7.118e-02 6.178e-02 1.152 0.24976
y.dl8 2.679e-02 4.572e-02 0.586 0.55818
du 3.590e-02 1.089e-02 3.295 0.00105 **
dt -1.278e-03 6.554e-04 -1.950 0.05175 .
---
Signif. codes: 0 '***' 0.001 '**' 0.01 '*' 0.05 '.' 0.1 ' ' 1

Residual standard error: 0.02392 on 507 degrees of freedom
(9 observations deleted due to missingness)
Multiple R-squared: 0.07668, Adjusted R-squared: 0.05483
F-statistic: 3.509 on 12 and 507 DF, p-value: 5.199e-05

Teststatistic: -8.393
Critical values: 0.01= -5.57 0.05= -5.08 0.1= -4.82

Potential break point at position: 503
```

for the CER price drivers:

$$y_t = \alpha + \beta_1 Brent_t + \beta_2 Ngas_t + \beta_3 Coal_t + \beta_4 Momentum_t + \beta_5 Linking_t + \varepsilon_t \quad (4.24)$$

with β_1 to β_3 the coefficients capturing the influence of energy prices in logre-turns transformation: *BRENTRET*, *NGASRET*, and *COALRET*. The brent price (expressed in Euro/BBL) is the brent crude futures Month Ahead price negoti-ated on ICE. The natural gas price used (expressed in Euro/Therm) is the futures Month Ahead natural gas price negotiated on Zeebrugge Hub. The coal price is the

Table 4.15 Cross-correlations between CER price drivers in logreturns transformation

	DIFFCER	DIFFBRENT	DIFFGAS	DIFFCOAL	MOMENTUM
DIFFCER	1				
DIFFBRENT	0.0712	1			
DIFFGAS	0.0266	0.0975	1		
DIFFCOAL	−0.0877	−0.0792	0.0321	1	
MOMENTUM	0.4661	−0.0084	−0.0148	0.0227	1

Antwerp/Rotterdam/Amsterdam (ARA, expressed in EUR/ton) coal futures Month Ahead price. The $Momentum_t$ variable, which is obtained as the difference between the CER variable at time t and at time $t − 5$, indicates bullish or bearish CER market trends (see [5]). $Linking_t$ is a dummy variable equal to one on the day of news announcements concerning the linking of emissions trading schemes worldwide, and zero otherwise. The latter variable indicates the ability of the CER to act as a 'world' carbon price, as regional and international emissions trading schemes establish linking procedures to recognize the validity of allowance trading between different systems. From that perspective, the CER price could act as the most fungible 'money' at the world level.

In Table 4.15, we check in the matrix of cross-correlations that there is no potential threat of multicollinearities[7] when estimating Eq. (4.24).

Table 4.16 allows to estimate Eq. (4.24) in R. Lines 1 to 11 load the relevant data. Line 12 conducts the estimation of the model.

Line 13 produces the estimation results contained in Table 4.17. We observe that $Brent_t$ is significant at the 10% level and positive. Therefore, more fuel use is also expected to increase the demand for CERs (in order to cover CO_2 emissions for companies covered by the emissions trading scheme). $Coal_t$ is also statistically significant (at the 5% level), and its sign is negative. The latter result is conform to the relationship discussed for EUAs: as the relative price of coal increases, companies need to rely less on this carbon-intensive energy source. Hence, the impact on the CER price is expected to be negative. Finally, the $Momentum_t$ is highly significant at the 1% level. Its positive sign may be explained by the fact that CER price changes have responded positively to carbon market trends during our study period. Other variables have not been found significant. Overall, we may conclude that energy prices have an influence on the CER price path (as for EUAs), while market-specific variables ($Momentum_t$) also seem to play a role in price formation (see [15] for more details).

Lines 14 and 15 give various diagnostic plots for residual analysis, which are shown in Fig. 4.10.

Following Lines 16 and 17, the ACF and PACF functions of the residuals are given in Fig. 4.11. Together, these diagnostic tools confirm visually that the residuals

---

[7]Note the dummy variable $Linking_t$ is not included.

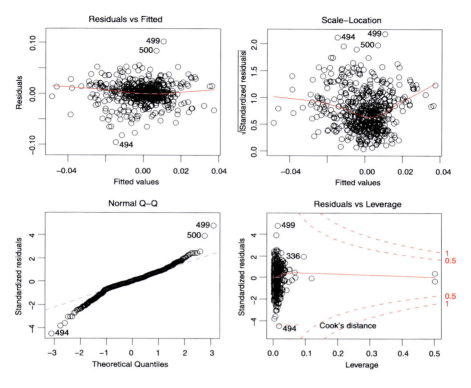

Fig. 4.10 Diagnostic plots for the model estimated in Eq. (4.24)

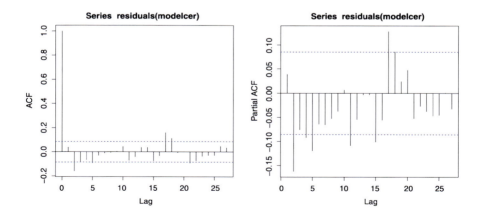

Fig. 4.11 Autocorrelation and partial-autocorrelation functions for the residuals from the model estimated in Eq. (4.11)

Table 4.16 R Code: Linear Regression Model for CERs

```
1 path<-"C:/"
2 setwd(path)
3 library(stats)
4 data=read.csv("data_chapter4.csv",sep=",")
5 attach(data)
6 cer=data[,3]
7 brent=data[,4]
8 gas=data[,5]
9 coal=data[,6]
10 momentum=data[,8]
11 linking=data[,9]
12 modelcer<-lm(cer~brent+gas+coal+momentum+linking)
13 summary(modelcer)
14 layout(matrix(1:4,2,2))
15 plot(modelcer)
16 acf(residuals(modelcer))
17 acf(residuals(modelcer),type='partial')
```

Table 4.17 R Output: Linear Regression Model for CERs

```
Call:
lm(formula = cer ~ brent + gas + coal + momentum + linking)

Residuals:
Min 1Q Median 3Q Max
-0.095802 -0.009498 0.000089 0.012281 0.101870

Coefficients:
Estimate Std. Error t value Pr(>|t|)
(Intercept) 0.0002007 0.0009400 0.214 0.8310
brent 0.0540701 0.0319227 1.694 0.0909 .
gas 0.0165778 0.0211729 0.783 0.4340
coal -0.1004529 0.0408671 -2.458 0.0143 *
momentum 0.0134956 0.0011017 12.250 <2e-16 ***
linking -0.0041577 0.0153401 -0.271 0.7865
---
Signif. codes: 0 '***' 0.001 '**' 0.01 '*' 0.05 '.' 0.1 ' ' 1

Residual standard error: 0.02156 on 523 degrees of freedom
Multiple R-squared: 0.2326, Adjusted R-squared: 0.2253
F-statistic: 31.7 on 5 and 523 DF, p-value: < 2.2e-16
```

of Eq. (4.24) are not autocorrelated, and thus the validity of the specification estimated.

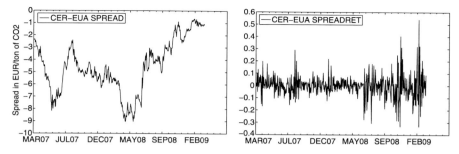

Fig. 4.12 CER-EUA Spread: Raw data (*left panel*) and logreturns (*right panel*) from March 09, 2007 to March 31, 2009

In the next section, we explore the reason for the price differences that exist between EUAs and CERs.

4.4 Arbitrage Strategies: The CER-EUA Spread

In this section, we explain the main arbitrage strategies by market operators behind the existence of the CER-EUA spread. In addition, we detail the main spread drivers.

4.4.1 Why So Much Interest in This Spread?

The CER-EUA spread is the premium between CERs and EUAs:

$$Spread_{t,i} = CER_{t,i} - EUA_{t,i} \qquad (4.25)$$

with $CER_{t,i}$ and $EUA_{t,i}$ the prices of, respectively, the CER and EUA futures contracts of maturity i, traded at time t in Euro on the ECX. Utilities trading desks on emissions trading pay a close attention to this spread, because they may benefit from 'free lunch' arbitrage opportunities simply by using secondary CER credits for their compliance under the EU ETS up to 13.4% on average. The arbitrage activity consists in buying secondary CERs and selling EUAs to cover their compliance position. For instance, a value of the spread equal to −3 Euro means that CER futures prices trade at 3 Euro less than the EUA futures price (for contracts of corresponding maturities). Traders closely watch this spread, since it represents the cost of swapping EUAs for CERs for use in meeting compliance targets under the EU ETS.

As we can see in Fig. 4.12, the value of the spread has been oscillating from −1 Euro to −9 Euro during the period under consideration. Significant arbitrage opportunities therefore exist for market agents who are able to surrender secondary CERs instead of EUAs in their own registries. Besides, in Table 4.18, we note that the mean value of the spread is equal to −4.54 Euro during our study period.

Table 4.18 Descriptive statistics for the CER-EUA Spread in raw (SPREAD) and returns (SPREADRET) transformation

	SPREAD	SPREADRET
Mean	−4.5459	−0.0013
Median	−4.6200	−0.0026
Maximum	−0.6478	0.5392
Minimum	−9.0435	−0.3329
Standard Deviation	2.1084	0.0876
Skewness	−0.0477	0.5256
Kurtosis	2.2923	8.4396
Jarque-Bera (JB)	11.2377	675.2977
Probability (JB)	0.0036	0.0001
Observations	529	528

Source: ECX

On financial markets, 'free lunch' arbitrage opportunities are supposed to be immediately executed by market agents in order to clear the pricing system. Since secondary CERs are essentially risk-free carbon credits (which have already been validated by the CDM Executive Board), their price should be equal to EUAs. Hence, we may ask the question: why do we observe a spread? The answer to this question is not as straightforward as it seems, and needs to be carefully explained through statistical analysis.

4.4.2 Spread Drivers

The CER-EUA spread may be explained by microstructure variables (see [14] for a survey). Since the arbitrage strategy consists in buying CERs and selling EUAs when the price difference between the two assets is maximum, we should be able to observe visually this effect in the volumes exchanged.

Figure 4.13 shows that when the CER-EUA spread is high (during the period of May 2008 when it is in the range of 8 Euro for instance), the volumes of EUAs and CERs exchanged are high as well. The main reason is that trading agents need to simultaneously trade both assets in order to benefit from the arbitrage opportunities. However, they do not always take advantage of the CER-EUA spread because the amount of CERs than can be used for compliance in EU ETS registries is limited to 13.4% on average among the 27 Member States. Therefore, there is a need to wait for the largest spread in order to maximize the potential source of revenue arising from this activity.

Note that the ability to benefit from the CER-EUA spread is limited to market agents which have their own registry and their own trading desks, such as power companies with dedicated trading infrastructures. Purely financial market players

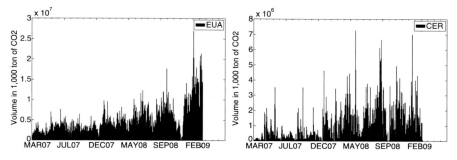

Fig. 4.13 CER-EUA Spread drivers: Trading volumes (in 1,000 ton of CO_2) for EUAs (*left panel*) and CERs (*right panel*)

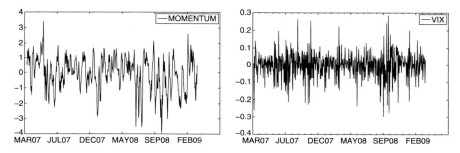

Fig. 4.14 CER-EUA Spread drivers: Market Momentum (*left panel*) and VIX Index (*right panel*)

(i.e. without access to a registry of CO_2 allowances linked to actual pollution at the plant level) can buy CERs and sell EUAs on exchanges, but they cannot surrender CERs for compliance in any registry.

Besides, we may remark that the volumes exchanged for both assets increase regularly, as the sphere of carbon markets is developing at the international level.

In Fig. 4.14, we show that the CER-EUA spread is not only linked to microstructure variables, but also to more fundamental drivers. Both the carbon market momentum (i.e. the degree to which a market enters bullish or bearish conditions) and the VIX Index (which is widely recognized as an indicator of aggregate market volatility, see [5]) can be shown as impacting the CER-EUA spread at statistically significant levels during our study period (see [15] for more details).

Appendix: Markov Regime-Switching Modeling with EUAs and CERs

We recall briefly here the basic framework behind Markov-switching models, as it has already been introduced in Chap. 3.

Consider an n-dimensional vector $\mathbf{y}_t \equiv (y_{1t}, \ldots, y_{nt})'$ which is assumed to follow a VAR(p) with parameters:

$$\mathbf{y}_t = \boldsymbol{\mu}(s_t) + \sum_{i=1}^{p} \boldsymbol{\phi}_i(s_t) y_{t-i} + \varepsilon_t \tag{4.26}$$

$$\boldsymbol{\varepsilon_t} \sim \mathcal{N}\left(\mathbf{0}, \sum(s_t)\right) \tag{4.27}$$

where the parameters for the conditional expectation $\boldsymbol{\mu}(s_t)$ and $\boldsymbol{\phi}_i(s_t)$, $i = 1, \ldots, p$, as well as the variances and covariances of the error terms ε_t in the matrix $\sum(s_t)$ all depend upon the state variable s_t which can assume a number q of values (corresponding to different regimes).

The model is estimated based on Gaussian maximum likelihood with $S_t = 1, 2$. By setting $S = [1\ 1]$, both the autoregressive coefficients and the model's variance are switching according to the transition probabilities.

Next, we apply this modelling framework to the time series of EUAs and CERs from March 2007 to March 2009. Thus, we obtain a two-regime Markov-switching VAR.

The estimation results are shown in Table 4.19. The statistically significant coefficients of the two means μ show the presence of switches between high-/low-growth periods. During expansion, output growth per month is equal to 0.03% on average. The time series is likely to remain in the expansionary phase with an estimated probability equal to 91%. Regime 1 is assumed to last 11 days on average. During recession, the average growth rate is equal to -0.05%. The probability that it will stay in recession is equal to 80%. The average duration of Regime 2 is equal to 5 days. According to the ergodic probabilities, the time series would spend 60% (40%) of the time spanned by the data sample in Regime 1 (Regime 2).

Interestingly, other coefficient estimates suggest that EUAs have several statistically significant effects on CERs: during Regime 1 (as $\phi_1 = 0.19$ and $\phi_2 = 0.07$ are significant at the 5% level), and during Regime 2 (as $\phi_1 = -0.39$ and $\phi_2 = 0.16$ are significant at the 1% level). Therefore, these results confirm the insights by Chevallier [3, 4] concerning the significant impact of EUAs on CERs, since the EU ETS is the most developed emissions market in the world to date. Concerning CERs, we can notice that they impact EUAs both during expansionary (as $\phi_1 = 0.70$ is significant at the 1% level) and recessionary (as $\phi_1 = -0.50$ is significant at the 5% level) phases.

The smoothed and regime transition probabilities are shown in, respectively, Figs. 4.15 and 4.16. They reveal essentially two states in the carbon futures markets: low-growth periods during April 2007 and April–December 2008, and high-growth periods during May 2007–May 2008 and January 2009–June 2009. These main switches from one regime to another seem to occur during compliance events or major changes in the underlying business cycle (with the entry of world economies into the recession in 2008), which affect the trading of CO_2 assets.

The diagnostic tests are shown in Table 4.20. The upper panel reports the results of three diagnostic tests. The first is a test of the Markov-switching model against

Table 4.19 Estimation results of the two-regime Markov-switching VAR for EUAs and CERs

Log-likelihood	251.13	
μ (Regime 1)	0.0003^{***}	
	(0.0001)	
μ (Regime 2)	-0.0005^{***}	
	(0.0001)	
Equation for *EUAFUTRET*	*EUAFUTRET*	*CERFUTRET*
ϕ_1 (Regime 1)	0.3320^{***}	0.7068^{***}
	(0.1124)	(0.1637)
ϕ_1 (Regime 2)	-0.5078^{***}	-0.5038^{**}
	(0.1969)	(0.2309)
ϕ_2 (Regime 1)	0.1187^{***}	0.1535
	(0.0113)	(0.1315)
ϕ_2 (Regime 2)	-0.2804	-0.4492^{*}
	(0.2061)	(0.2508)
Equation for *CERFUTRET*	*EUAFUTRET*	*CERFUTRET*
ϕ_1 (Regime 1)	0.1975^{**}	0.4187^{***}
	(0.0830)	(0.1352)
ϕ_1 (Regime 2)	-0.3963^{***}	-0.4997^{***}
	(0.1536)	(0.1959)
ϕ_2 (Regime 1)	0.0746^{**}	0.1269
	(0.0304)	(0.1131)
ϕ_2 (Regime 2)	0.1654^{***}	-0.2053
	(0.0541)	(0.1805)
Standard error (Regime 1)	0.0004	
Standard error (Regime 2)	0.0005	
Transition Probabilities Matrix	Regime 1	Regime 2
Regime 1	0.9104^{***}	0.2047
	(0.0662)	(0.1405)
Regime 2	0.0953	0.8036^{***}
	(0.0617)	(0.1880)
Regime Properties	Prob.	Duration
Regime 1	60.63	11.30
Regime 2	39.37	4.92

Note: *EUAFUTRET* stands for the EUA Futures Price in Logreturn form. *CERFUTRET* stands for the CER Futures Price in Logreturn form. Standard errors are in parentheses ***, **, * denote respectively statistical significance at the 1%, 5% and 10% levels

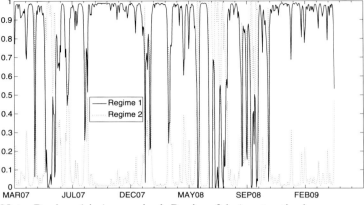

Note: Regime 1 is 'expansion'. Regime 2 is 'contraction'.

Fig. 4.15 Smoothed transition probabilities estimated from the two-regime Markov-switching VAR for EUAs and CERs

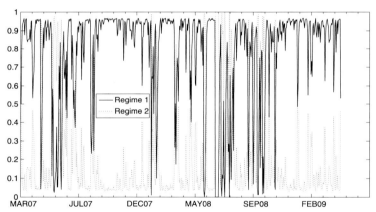

Note: Regime 1 is 'expansion'. Regime 2 is 'contraction'.

Fig. 4.16 Regime transition probabilities estimated from the two-regime Markov-switching VAR for EUAs and CERs

the simple nested null hypothesis that the data follow a geometric random walk with i.i.d. innovations. Note M the p-value from the LR test:

$$\mathrm{Prob}[LR(q^*)] > M = \mathrm{Prob}(\chi_d^2 > M) + \frac{V M^{(d-1)/2} e^{-M/2} 2^{-d/2}}{\Gamma(d/2)} \qquad (4.28)$$

where $\mathrm{Prob}(LR(q^*) > M | H_0)$ is the upper bound critical value, LR is the likelihood ratio statistic, q^* is the vector of transition probabilities ($q^* = \arg\max LnL(q) | H_1$) and d is the number of restrictions under the null hypothesis.

Table 4.20 Diagnostic tests of the two-regime Markov-switching VAR for EUAs and CERs

	Markov-switching VAR	
LR Statistic	25.981	
p-value	0.001	
Symmetry test	2.069	
p-value	0.022	
RCM 2-State	38.6110	
Distributional Characteristics	*EUAFUTRET*	*CERFUTRET*
Mean	0.0011	0.0006
Median	0.0014	0.0004
Maximum	1.1250	1.1272
Minimum	−2.5437	−1.1714
Std. Dev.	0.3835	0.2193
Skewness	−0.4556	−0.3944
Kurtosis	6.4737	5.4327

Note: Distributional characteristics are given for the Markov-switching processes implied by the estimates in Table 4.19. *EUAFUTRET* stands for the EUA Futures Price in Logreturn form. *CERFUTRET* stands for the CER Futures Price in Logreturn form

In Table 4.20, this adjustment produces a LR statistic equal to 25.98. We reject the random walk at the 0.1 percent level. We conclude that the relationship is better described by a two-regime Markov-switching model than by the random walk model.

The second test reported in Table 4.20 is for symmetry of the Markov transition matrix, which implies symmetry of the unconditional distribution of the growth rates. This test examines the maintained hypothesis that p (the probability of being in a high-growth state or boom) equals q (the probability of being in a low-growth state or depression) against the alternative that $p < q$. Table 4.20 reports statistics that are asymptotically standard normal under the null. We reject the hypothesis of symmetry at the 5% level.

Third, Ang and Bekaert (2002, [1]) set out a formal definition of and a test for regime classification. They argue that a good regime switching model should be able to classify regimes sharply. Weak regime inference implies that the regime-switching model cannot successfully distinguish between regimes from the behavior of the data, and may indicate misspecification. To measure the quality of regime classification, we therefore use Ang and Bekaert's (2002 [1]) Regime Classification Measure (RCM) defined for two states as:

$$RCM = 400 \times \frac{1}{T} \sum_{t=1}^{T} p_t (1 - p_t) \tag{4.29}$$

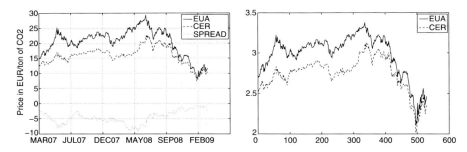

Fig. 4.17 Raw time-series (*left panel*) and natural logarithms (*right panel*) of the ECX EUA December 2008/2009 Futures and Reuters CER Price Index from March 9, 2007 to March 31, 2009

where the constant serves to normalize the statistic to be between 0 and 100, and p_t denotes the ex-post smoothed regime probabilities. Good regime classification is associated with low RCM statistic values. A value of 0 indicates that the two-regime model is able to perfectly discriminate between regimes, whereas a value of 100 indicates that the two-regime model simply assigns each regime a 50% chance of occurrence throughout the sample. Consequently, a value of 50 is often used as a benchmark (see Chan et al. (2011, [2]) for instance).

Adopting this definition to the current context, the RCM 2-State statistic is equal to 38.61 in Table 4.20. It is substantially below 50, consistent with the existence of two regimes. It is very interesting that our estimated Markov-switching model has classified the two regimes extremely well.

Finally, Table 4.20 reports the distributional characteristics for the Markov-switching processes implied by the estimates in Table 4.19. We conclude that the Markov-switching model produces both the degree of skewness and the amount of kurtosis that are present in the original data.

Overall, the Markov-switching modelling brings us new insights as to when market shocks occur, and actually impact the EUA and CER futures price series. The results show that significant interactions exist between the two markets, especially during periods of economic recession when the market trends are destabilised.

Problems

Problem 4.1 (The EUA-CER Spread)

(a) Consider the time-series in Fig. 4.17: briefly describe the variables that you observe, as well as their behaviour. What are the differences between the two panels?

(b) Consider the descriptive statistics in Table 4.21: briefly comment on the descriptive statistics that you observe for each variable.

In what follows, the ECX EUA December 2008/2009 and Reuters CER Price Index Raw Series are expressed in *natural logarithms*.

Table 4.21 Summary Statistics for the Raw Time Series and Natural Logarithms

Variable	Mean	Median	Max	Min	Std. Dev.	Skew.	Kurt.
Raw Price Series							
EUA	20.4038	21.5200	29.3300	8.2000	4.4592	−0.7659	3.0319
CER	15.8579	16.0687	22.8500	7.4846	2.9864	−0.3514	3.1352
Natural Logarithms							
EUA	2.9866	3.0689	3.3786	2.1041	0.2551	−1.3231	4.2758
CER	2.7439	2.7768	3.1289	2.0128	0.2055	−0.9947	4.1821

Note: *EUA* refers to ECX EUA December 2008/2009 Futures, and *CER* to Reuters CER Price Index. *Std. Dev.* stands for Standard Deviation, *Skew.* for Skewness, and *Kurt.* for Kurtosis. The number of observations is 529

Table 4.22 Cointegration Rank: Maximum Eigenvalue Statistic

Hypothesis	Statistic	10%	5%	1%
$r \leq 1$	1.65	6.50	8.18	11.65
$r = 0$	6.81	12.91	14.90	19.19

Table 4.23 Cointegration Rank: Trace Statistic

Hypothesis	Statistic	10%	5%	1%
$r \leq 1$	1.65	6.50	8.18	11.65
$r = 0$	8.46	15.66	17.95	23.52

Table 4.24 Cointegration Vector

Variable	EUA(1)	CER(1)
EUA(1)	1.0000	1.0000
CER(1)	−1.291699	0.2616575

Note: *EUA* refers to ECX EUA December 2008/2009 Futures, and *CER* to Reuters CER Price Index, transformed to natural logarithms. Lag order in parenthesis. The number of observations is 529

(c) Comment on Tables 4.22 and 4.23: which cointegration space can you identify, given a 5% significance level? Provide detailed comments as to which statistics you use, and how you to read these tables.

(d) Comment on Tables 4.24, 4.25 and 4.26: based on your conclusion in Question (c), interpret the following Error Correction Model.

In what follows, the ECX EUA December 2008/2009 and Reuters CER Price Index Raw Series are expressed in *log-returns*.

Table 4.25 Model Weights

Variable	EUA(1)	CER(1)
EUA.d	0.01979082	−0.003906266
CER.d	0.02820092	−0.001641289

Note: *EUA* refers to ECX EUA December 2008/2009 Futures, and *CER* to Reuters CER Price Index, transformed to natural logarithms. Lag order in parenthesis. *.d* refers to difference. The number of observations is 529

Table 4.26 VECM with $r = 1$

Variable	EUA.d	CER.d
Error Correction Term *ect*	−0.0197908	−0.0282009
Deterministic constant	0.0106349	0.0154190
Lagged differences		
EUA.d(1)	−0.0641515	−0.0504123
CER.d(1)	0.2307197	0.1423340

Note: *EUA* refers to ECX EUA December 2008/2009 Futures, and *CER* to Reuters CER Price Index, transformed to natural logarithms. Lag order in parenthesis. *.d* refers to difference. *ect* refers to the Error Correction Term. The number of observations is 529

Fig. 4.18 Logreturns of the ECX EUA December 2008/2009 Futures and Reuters CER Price Index from March 9, 2007 to March 31, 2009

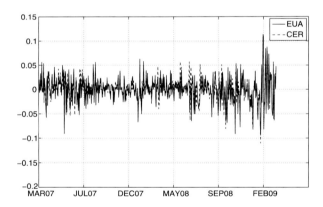

(e) Consider the time-series in Fig. 4.18: briefly describe the variables that you observe, and their behaviour. What are the differences compared to Fig. 4.17?

(f) Consider the descriptive statistics in Table 4.27: briefly comment on the descriptive statistics that you observe for each variable.

  You are given the following information:

• Optimal lag selection: $AIC(n) = 4$, $HQ(n) = 1$, $SC(n) = 1$, $FPE(n) = 4$.
• p-value of the Portmanteau test $< 5\%$ for VAR(1) residuals.

Table 4.27 Summary Statistics for the Log-Returns

Variable	Mean	Median	Max	Min	Std. Dev.	Skew.	Kurt.
Log-Returns							
EUA	−0.0004	0.0001	0.1136	−0.0943	0.0268	−0.0608	4.8680
CER	−0.0003	0.0001	0.1125	−0.1104	0.0244	−0.3703	5.9619

Note: *EUA* refers to ECX EUA December 2008/2009 Futures, and *CER* to Reuters CER Price Index. *Std.Dev.* stands for Standard Deviation, *Skew.* for Skewness, and *Kurt.* for Kurtosis. The number of observations is 529

Table 4.28 VAR(4) Model Results for the EUA Variable

Variable	Estimate	Std. Error	t-value	$\Pr(> |t|)$
Lagged levels				
EUA(1)	−0.06061	0.08233	−0.736	0.46193
CER(1)	0.2454***	0.08952	2.741	0.00634
EUA(2)	−0.1504*	0.08227	−1.828	0.06806
CER(2)	0.01805	0.08981	0.201	0.84082
EUA(3)	−0.1275	0.08234	−1.549	0.12198
CER(3)	0.2056**	0.08992	2.287	0.02261
EUA(4)	0.152601*	0.08203	1.860	0.06342
CER(4)	−0.07958	−0.09050	−0.879	0.37965
Deterministic				
const.	0.003480	0.002344	1.484	0.13831
trend	−0.000015**	0.000008	−1.912	0.05646
Residuals Std. Error	0.02613			
R-Squared	0.07136			
Adjusted R-Squared	0.0551			
F-Statistic	0.00001			

Note: *EUA* refers to ECX EUA December 2008/2009 Futures, and *CER* to Reuters CER Price Index. Lag order in parenthesis. *Std. Error* stands for Standard Error, *const.* for deterministic constant, and *trend* for deterministic trend. *p*-value is given for the *F*-Statistic ***denotes 1%, **5%, and *1% significance levels. The number of observations is 529

(g) Comment on Tables 4.28 and 4.29: provide detailed comments of the static VAR analysis for both variables.

(h) Comment on Table 4.30: analyze the diagnostic tests for the VAR(4) model. Can you conclude that the model is well-specified?

(i) Comment on Table 4.31: what can you conclude concerning the Granger causality tests?

(j) Consider the time-series in Fig. 4.19: briefly comment on the impulse response functions that you observe for each variable. Given your analysis in Questions

Table 4.29 VAR(4) Model Results for the CER Variable

Variable	Estimate	Std. Error	t-value	$Pr(> |t|)$
Lagged levels				
EUA(1)	−0.05705	0.07546	−0.756	0.44995
CER(1)	0.1627**	0.08206	1.983	0.04787
EUA(2)	−0.08439	0.07541	−1.119	0.26360
CER(2)	−0.01538	0.08232	−0.187	0.85188
EUA(3)	−0.1937***	0.07547	−2.566	0.01056
CER(3)	0.2476***	0.08242	3.004	0.00279
EUA(4)	0.1201	0.07519	1.597	0.11079
CER(4)	0.001684	0.08296	0.020	0.98381
Deterministic				
const.	0.002553	0.002149	1.188	0.23532
trend	−0.00001	0.000007	−1.517	0.12979
Residuals Std. Error	0.02395			
R-Squared	0.06089			
Adjusted R-Squared	0.04445			
F-Statistic	0.00016			

Note: *EUA* refers to ECX EUA December 2008/2009 Futures, and *CER* to Reuters CER Price Index. Lag order in parenthesis. *Std. Error* stands for Standard Error, *const.* for deterministic constant, and *trend* for deterministic trend. The number of observations is 529

Table 4.30 Diagnostic tests of VAR(4) Model

Test	Statistic	D.F.	p-value
Portmanteau	57.4878	48	0.16
ARCH VAR	97.1946	9	0.01
JB VAR	147.6817	4	0.01
Kurtosis	143.5005	2	0.01
Skewness	4.1811	2	0.1236

Note: Portmanteau is the asymptotic Portmanteau test with a maximum lag of 16, ARCH VAR is the multivariate ARCH test with a maximum lag of order 5, JB is the Jarque Berra normality test for multivariate series applied to the residuals of the VAR(4), Kurtosis and Skewness are separate tests for multivariate skewness and kurtosis

(c) to (i), which Cholesky decomposition has been chosen to compute these graphs?

In what follows, the GARCH(p, q) model with a Gaussian conditional probability distribution has been estimated by Quasi-Maximum Likelihood with the BHHH algorithm.

Table 4.31 Granger
Causality Tests

Test	Statistic	p-value
Cause = CER		
Granger	3.2237	0.01215
Instant	218.49	0.0001
Cause = EUA		
Granger	2.7535	0.02695
Instant	218.49	0.0001

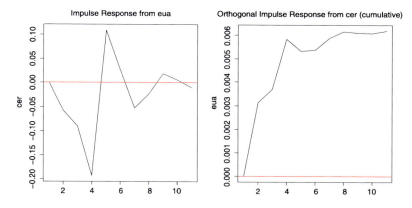

Fig. 4.19 Impulse responses from EUA to CER (*left panel*) and from CER to EUA (*right panel*)
Variables of the VAR(4) Model

Fig. 4.20 GARCH(1, 1)
model autocorrelation of
squared logreturns of the
CER-EUA Spread

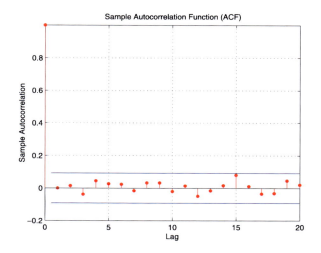

Table 4.32 GARCH estimates for the CER-EUA Spread

Mean Equation	
α	-0.0002
	(0.0006)
Variance Equation	
κ	0.0002^{***}
	(0.0000)
β	0.1110
	(0.1207)
ϕ	0.3194^{***}
	(0.0504)
Diagnostic Tests	
Ljung-Box	0.9631
ARCH test	0.9687
Log-Likelihood	1331.7
AIC	-2619.3
BIC	-2602.8

Note: The values reported in parentheses are the standard errors. ***indicates 1% level significance, **5% level significance, and *1% level significance[8]

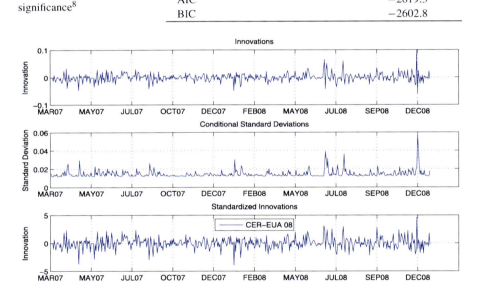

Fig. 4.21 GARCH(1, 1) model of the CER-EUA Spread innovations (*top panel*), conditional standard deviations (*middle panel*) and standardized innovations (*bottom panel*)

(k) Comment on Table 4.32: provide detailed comments for *each parameter* of the GARCH(p, q) model.

---

[8]The model estimated is:

$$spread = \alpha + \varepsilon_t \tag{4.30}$$
$$\sigma_t^2 = \kappa + \beta\sigma_{t-1}^2 + \phi\varepsilon_{t-1}^2 \tag{4.31}$$

(l) Comment on Fig. 4.20: what can you conclude with respect to the autocorrelation structure of GARCH estimates?

(m) Consider the time series in Fig. 4.21: as a diagnostic check of these estimates, what can you conclude from Fig. 4.21?

References

1. Ang A, Bekaert G (2002) Regime switches in interest rates. J Bus Econ Stat 20:163–182
2. Chan KF, Treepongkaruna S, Brooks R, Gray S (2011) Asset market linkages: evidence from financial, commodity and real estate assets. J Bank Finance 35:1415–1426
3. Chevallier J (2010) EUAs and CERs: vector autoregression, impulse response function and cointegration analysis. Econ Bull 30:558–576
4. Chevallier J (2011) Anticipating correlations between EUAs and CERs: a dynamic conditional correlation GARCH model. Econ Bull 31:255–272
5. Collin-Dufresne P, Goldstein RS, Spencer Martin J (2001) The determinants of credit spread changes. J Finance 56:2177–2202
6. Engle R (1982) Autoregressive conditional heteroskedasticity with estimates of the variance of United Kingdom inflation. Econometrica 50:987–1007
7. Hamilton JD (1996) Time series analysis, 2nd edn. Princeton University Press, Princeton
8. Johansen S (1988) Statistical analysis of cointegration vectors. J Econ Dyn Control 12:231–254
9. Johansen S, Juselius K (1990) Maximum likelihood estimation and the demand for money inference on cointegration with application. Oxf Bull Econ Stat 52:169–210
10. Johansen S (1991) Estimation and hypothesis testing of cointegration vectors in Gaussian vector autoregressive models. Econometrica 59:1551–1580
11. Johansen S (1992) Cointegration in partial systems and the efficiency of single-equation analysis. J Econom 52:389–402
12. Lütkepohl H, Saikkonen P, Trenkler C (2004) Testing for the cointegrating rank of a VAR with level shift an unknown time. Econometrica 72:647–662
13. Lütkepohl H (2006) New introduction to multiple time series analysis. Springer, New York/Dordrecht/Heidelberg/London
14. Madhavan A (2000) Market microstructure: a survey. J Financ Mark 3:205–258
15. Mansanet-Bataller M, Chevallier J, Herve-Mignucci M, Alberola A (2011) EUA and sCER phase II price drivers: unveiling the reasons for the existence of the EUA-sCER spread. Energy Policy 39:1056–1069
16. Perron P (1989) The great crash, the oil price shock, and the unit root hypothesis. Econometrica 57:1361–1401
17. Pfaff B (2008) Analysis of integrated and cointegrated time series with R. Springer, New York/Dordrecht/Heidelberg/London
18. Pinho C, Madaleno M (2011) CO_2 emission allowances and other fuel markets interactions. Environ Econ Policy Stud 13:259–281
19. Ploberger W, Kramer W (1992) The CUSUM test with OLS residuals. Econometrica 60:271–285
20. Sims CA (1980) Macroeconomics and reality. Econometrica 48:1–48
21. Trenkler C (2003) A new set of critical values for systems cointegration tests with a prior adjustment for deterministic terms. Econ Bull 3:1–9
22. Zivot E, Andrews DWK (1992) Further evidence on the great crash, the oil-price shock, and the unit-root hypothesis. J Bus Econ Stat 10:251–270

Chapter 5
Risk-Hedging Strategies and Portfolio Management

Abstract This chapter considers first the risk factors impacting jointly the Clean Development Mechanism and the European Union Emissions Trading Scheme. Second, it provides an econometric modelling exercise of risk premia in CO_2 allowance spot and futures prices under the EU ETS (based on Bessembinder and Lemmon, J. Finance 57:1347–1382, 2002), which takes into account the specificities of carbon allowances in terms of storage. Third, carbon price risks in the power sector are identified econometrically by analyzing the factors that influence fuel-switching. Fourth, a stylized portfolio management exercise is proposed by using mean-variance optimization in a broadly diversified portfolio composed of stocks, bonds, weather, energy and carbon assets. The Appendix details how to compute implied volatility series from options, which can then be used to represent the risk-neutral probability distribution of carbon prices.

5.1 Risk Factors

Risk management constitutes an important topic for companies falling under the regulation of an emissions trading system. Indeed, following the creation of an ETS, the price variation of CO_2 allowances constitutes a new source of risk against which utilities need to hedge, and to devise appropriate risk management strategies. Blyth et al. (2009, [8]) and Blyth and Bunn (2011, [9]) underline that price formation in carbon markets involves a complex interplay between policy targets, dynamic technology costs, and market rukes. Besides, they note that policy uncertainty is a major source of carbon price risk. In a regulatory context with more certainty, there would be benefits to transfer risk mitigation from governments to market participants' business risk.

In order to better understand these various sources of risk on the carbon market, we review first the risk factors on the European Union Emissions Trading Scheme (EU ETS) and the Clean Development Mechanism (CDM). Concerning the risk characteristics of CO_2 allowances, we detail first the idiosyncratic risks affecting each emissions market, be it in terms of regulatory uncertainty or economic factors. Second, we assess common risk factors between CERs and EUAs by focusing more particularly on the role played by the CER import limit within the EU ETS.

5.1.1 Idiosyncratic Risks

In this section, we examine separately the risk factors affecting the CDM and the European carbon market.

5.1.1.1 Risk Factors Specific to CERs

Under the Kyoto Protocol, projects from the CDM add value to an investment through Certified Emissions Reductions (CER) revenues. In the exchange contract of guaranteed secondary CER (sCERs), the seller agrees to pay EUA mark-to-market/liquidated damages CERs or cash in case of non-delivery.[1] CDM projects are registered by the CDM Executive Board (CDM EB) from the United Nations Framework Convention on Climate Change (UNFCCC), and displayed in the CDM pipeline[2] along the several steps from registration to validation.

The risk factors specific to CERs are as follows. On the primary market, there needs:

1. An increased predictability of issuance and frequency of transfer CERs by the CDM EB,
2. An International Transactions Log (ITL) operational and linked to EU registry system, and
3. The development of a robust options market.

On the secondary market, there needs to take into account:

1. The acceptance into EU registries,
2. The development of new emissions trading schemes, which increases demand and thus pushes CER prices to converge with those under compliance schemes, and
3. Limitations on the use of CERs compliance under the EU ETS.

Finally, CERs are affected by the uncertainties concerning the post-2012 climate regime.

5.1.1.2 Risk Factors Specific to EUAs

We may identify three main sources of risk on the EU ETS. First, the rents distributed to existing incumbents on a 'first-come, first-served' basis represent a market value of 40 billion Euro that was created at the same time as CO_2 emissions were capped. This allocation methodology is also known as grandfathering. Since

---

[1]Note CERs traded on exchanges are precisely guaranteed secondary CERs (sCERs), but for simplicity we use the common denomination term of CERs throughout this chapter.

[2]See http://www.cdmpipeline.org.

January 1, 2005 carbon allowances form another asset in commodities against which industrials and brokers need to hedge. As the volume of transaction on the EU ETS has been increasing steadily (from 262 million tons in 2005 to 4,414 million tons in 2010), this trading activity reflects market participants' progressive learning of this new financial market.

Second, during Phase I of the EU ETS (2005–2007), EUAs experienced a high level of volatility around each compliance event [2, 20]. Industrial installations have the obligation to surrender to the European Commission (EC) the exact number of allowances that matches their verified emissions each year around end of March.[3] The official report by the European Commission is disclosed by mid-May, but installation operators have already a fair amount of information between the publication of their own report and the compilation of verified emissions by the European Commission to approximate the global level of emissions relative to allowances allocated, and to adjust their anticipations. Previous literature has documented how structural breaks occur in EUA price series during the 2005 [2] and 2006 [20] compliance events.

Third, installations do not need to physically hold allowances to produce, but only to match them with verified emissions for their yearly compliance report to the European Commission. Consequently, the probability of a potential illiquidity trap exists if market participants face a market squeeze during the compliance event. Another specificity of emission allowances may be highlighted: compared to stocks which are valid during the entire lifetime of the firm, emission allowances are vintaged for a given compliance year and cannot be used for future compliance periods, unless intertemporal flexibility mechanisms are authorized. During Phase I of the EU ETS, the inter-period transfer of allowances has been restricted by all Member States [1]. There is now unlimited banking and borrowing between Phases II and III.

5.1.2   Common Risk Factors

Similarly, we may identify several common risk factors between EUAs and CERs.

5.1.2.1   The ITL-CITL Connection

On October 19, 2008 the European Commission's 'Community Independent Transaction Log' (CITL) connected to the United Nations' 'International Transaction Log' (ITL). The ITL-CITL connection involved the EU 27 Member States disconnecting their national registries, and re-connecting them to the ITL. Simultaneously, the CITL connected directly to the ITL. These operations allowed the delivery of sellers' international credits from the Kyoto Protocol (such as CERs) into buyers'

---

[3]See Chap. 1 for the detailed calendar.

EU national registries. Before that date, issued CERs remained into the UN CDM Registry, waiting for the connection with the EU registry to be completed.

During Phase II of the EU ETS (2008–2012), the EU Commission has announced that the use of CERs by industrial installations will be capped at 270 million tons per year. For Phase III, the limit for existing installations will be 40 million tons per year. All together, Phase III CER imports will add up to 300 million tons. Thus, the post-2012 EU ETS CER import limit has been fixed at 1.7 billion tons, coupled with quality criteria and eligibility requirements (namely to reduce the share of HFC-23 projects from China and India).[4]

Therefore, the ITL-CITL connection has filled the risk factor common to CDM project developers and investors. First, it has provided stimulus to CERs and EUAs trading, as the barriers to transfer allowances between UN and EU registries have been removed. Second, it has offered to EU investors in CDM projects the possibility to monetize their assets, by using CERs towards their own compliance (for regulated utilities having their own registry) or by exchanging them on the European market as secondary CERs (for pure financial players such as consulting firms in project development and investment banks). However, it seems worth noting that some types of CERs credits are restricted from use in the EU ETS, such as Land Use, Land Use Change and Forestry (LULUCF) projects.

5.1.2.2 The Import Limit of CERs Within the EU ETS

Since the ITL-CITL connection, Kyoto credits may be imported into the EU ETS, and are valid for compliance up to 13.4% on average. As detailed in Chap. 4, the EUA-CER price arbitrage becomes possible up to that limit. Investors will indeed maximize their profits by buying the maximum volume of CERs allowed for compliance and selling the same amount EUAs when the EUA-CER spread is at its maximum, i.e. in the range of 6 to 9 Euro per ton of CO_2.

Concerning the role played by the CER import limit for compliance within the EU trading system, firms cannot use more than 270 million tons of CERs for compliance per year (according to the EU Linking Directive) during Phase II. Besides, the path of CERs delivery critically depends on the criteria for CDM project validation. In terms of expected CER delivery, HFC (hydrofluorocarbon), PFC (perfluorocarbon), CH_4 (methane) and N_2O (nitrous oxide) emission reduction projects have a relatively large market share, which is mainly due to these gases' high global warming potential. It appears also important to look at the distribution of projects and technologies in terms of expected GHG emission reductions. According to the CDM Pipeline, projects involving HFC emissions reduction constitute 17% of expected emissions reductions[5] by 2012. Projects involving N_2O emissions reduction

---

[4]CDM credits can be issued by destroying HFC-23 (trifluoromethane), which is 11,700 times more potent as a greenhouse gas than CO_2. It can be removed by installing gas scrubbers for instance.

[5]Recall that each ton of HFC abated delivers 11,700 CERs.

constitute 9% of expected CERs.[6] By contrast, hydro projects will deliver 'only' 17% of expected CERs, although their number is much larger. Hence, the EC has enacted new measures in order to gradually phase out CDM projects originating from HFC-23 (and other gases) in order to keep incentives for further emissions reductions in the EU.

In terms of risk-assessment, it is also interesting to note that the delivery of credits, and thus the performance of projects, greatly varies: only 10 to 15% of methane reduction credits could be verified and therefore issued. This ratio is equal to 75% for HFC-23 emissions reduction projects, while N_2O projects sometimes deliver more CERs than expected. Of course, the list of common risk factors between the CDM and EU ETS markets would not be complete without a careful outlook of negotiations for a post-Kyoto treaty, with the next COP/MOP Meeting scheduled in Durban in December 2011 (see Chap. 1 for more details).

Having detailed the risk factors specific to emissions markets (CDM and EU ETS), we proceed in the next section with a more formal econometric analysis of risk premia in CO_2 spot and futures prices.

5.2   Risk Premia

Risk premia can be broadly defined as the evolution through time of agents' preferences driving volatilities of spot and futures prices [13]. This section analyzes the modelling of risk premia in CO_2 allowances spot and futures prices, valid for compliance under the EU ETS. First, we recall some theory on the relationships between spot and futures prices for commodity markets. Similarly to electricity markets, a salient characteristic of CO_2 allowances is that the theory of storage does not hold, as CO_2 allowances only exist on the balance sheets of companies regulated by the scheme. That is why we study then Bessembinder and Lemmon's (2002, [7]) model of the futures-spot bias, along with an empirical application.

5.2.1   Theory on Spot-Futures Relationships in Commodity Markets

Assuming rational expectations and risk-neutral market agents, future spot prices should only deviate from futures prices in case of unexpected shocks. Under such restrictive assumptions, spot prices in the delivery period S_T should equal futures prices $F_{t,T}$ plus a white noise error term ε_t with zero mean [12, 43]:

$$S_T = F_{t,T} + \varepsilon_t \tag{5.1}$$

For commodity markets, the difference between today's spot and futures price is usually based on the theory of storage. In addition to interest foregone through

---

[6]The N_2O global warming potential is 310 times higher than CO_2.

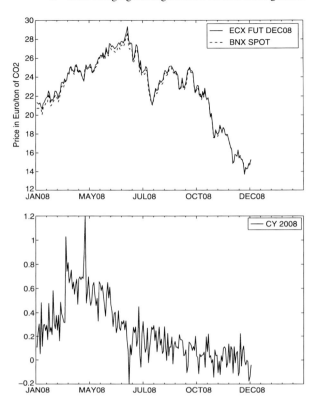

Fig. 5.1 Carbon Spot and Futures Prices (*top*) and convenience yield (*bottom*) from January 2 to December 15, 2008. Source: Bluenext, European Climate Exchange, Euribor

the commodity, storage costs for holding the commodity and a convenience yield on inventory have to be considered. Therefore, the interest rate, storage costs and convenience yield form a limit for futures prices [28, 43]:

$$F_{t,T} = S_t e^{(r+s)(T-t)} \tag{5.2}$$

where $F_{t,T}$ is the futures price at time t for delivery in T, S_t is the spot price at time t, r is a constant interest rate, and s are storage costs. If the futures price deviates from this relationship, then arbitrageurs would be able to make risk-free profits.

As Pindyck (2001, [37]) puts it, this no-arbitrage condition states that the only cost of buying a commodity at time t and delivering it at maturity T is the foregone interest. Agents incur the opportunity cost of purchasing the asset, but in return they benefit from possessing the commodity and being able to trade it until maturity.

Hence, the convenience yield at time t for maturity T may be modeled directly as:

$$y_{t,T} = S_t e^{r(T-t)} - F_{t,T} \tag{5.3}$$

As an illustrative example, the top panel of Fig. 5.1 pictures the carbon spot and futures prices for the contract of maturity December 2008, along with the corresponding convenience yield in the bottom panel.

In the bottom panel of Fig. 5.1, we may observe that the convenience yield is strongly time-varying during 2008, going from -0.2 in June to 1.2 in April. These variations may be explained by:

1. the delayed effect of the 'credit crunch' crisis on the carbon market (see Chevallier (2009, [16])), and
2. the 2007 compliance event which occurred in April 2008 (see Chevallier et al. (2009, [20]) for a detailed analysis of the effects of compliance events on investors' expectation changes).[7]

However, as storage costs for CO_2 allowances are null (they only exist in companies' balance sheets), the theory of storage does not appear as directly applicable for CO_2 allowances. As discussed by Chevallier (2010, [18]), CO_2 and power prices share the same characteristics in terms of storage (the costs of storage are almost infinite for electricity prices, while they are virtually inexistent for CO_2 allowances), i.e. the theory of storage does not apply for either of these two commodities. Besides, Chevallier et al. (2009, [20]) provide statistical evidence that the cost-of-carry relationship[8] does not hold between CO_2 spot and futures prices of varying delivery dates (see also [1]).

As shown in Eq. (5.1), futures prices can be related to the expected spot prices. Hence, we can use another approach than the theory of storage to determine equilibrium relationships for CO_2 prices, as explained in the next section.

5.2.2 Bessembinder and Lemmon's (2002) Futures-Spot Structural Model

According to [7] and [32], expected spot prices are built on fundamental expectations of market participants, and are translated to futures prices by applying risk premia which are a compensation for bearing the spot price risk. In the case of CO_2 allowances, the expectations and risk preferences of market participants determine futures prices. This assumption constitutes another approach for which the futures price at time t with maturity T, $F_{t,T}$ is split into the expected future spot price $E(S_T \mid \Omega_t)$ and a risk premium π_t^F. The latter represents a premium (discount) that buyers (sellers) of futures contracts are willing to pay (accept) in addition to the expected future spot price in order to eliminate the risk of unfavorable spot price

---

[7]For a more detailed time-series modeling of the convenience yield (especially by using intraday data) in the carbon market, see the analysis by Chevallier (2009, [17]).

[8]According to the cost-of-carry relationship, and without storage costs for carbon allowances, the futures and spot prices are linked through $S_t = F_T e^{r(Tt)}$ with S_t the spot price at time t, F_T the futures prices of a contract with delivery in T, and r the interest rate [12, 43].

movements [11, 21, 25]. From that perspective, the futures price can be computed as:

$$F_{t,T} = E(S_T \mid \Omega_t) + \pi_t^F \tag{5.4}$$

with Ω_t the information set available at time t. If risk premia exist, futures prices are not unbiased predictors of future spot prices.

As discussed above, we need to take into account in our empirical analysis the specificity of CO_2 allowances in terms of storage. Similarly to electricity markets where storage costs are very large and almost infinite [7], the theory of storage [28, 43] does not appear directly applicable for CO_2 allowances where storage costs are null. Unlike other commodity markets such as oil and gas, the cost of storage for CO_2 allowances is indeed virtually inexistent, as the costs incurred by detaining CO_2 allowances only appear in the balance sheets of utilities regulated by the EU ETS [18].

That is why we propose to study in this section Bessembinder and Lemmon's (2002, [7]) futures-spot structural model (which was first applied to the electricity market) to the EU ETS (as in [18]). In their model, the authors state that the futures premium (i.e. the difference between futures and expected spot prices) is a function of the variance and skewness of spot prices. The goal is to use the variance and skewness of the spot price distribution as a risk assessment tool of market participants. The futures price is determined as:

$$F_{t,T} = E(S_T) + \alpha \, \mathrm{Var}(S_T) + \gamma \, \mathrm{Skew}(S_T) \tag{5.5}$$

with $\mathrm{Var}(S_T)$ and $\mathrm{Skew}(S_T)$, respectively, the variance and unstandardized skewness of the spot price at maturity S_T. Since α is expected to be negative and γ positive, the risk premium is negatively related to the variance of the spot price, and positively related to the unstandardized skewness of the spot price.

We apply this methodology in the context of CO_2 allowances spot and futures prices in the next section.

5.2.3 *Empirical Application*

In order to test for the implications of Bessembinder and Lemmon's (2002, [7]) model for CO_2 allowances, we run the following regression:

$$F_{t,T} - S_T = c + \alpha \, \mathrm{Var}(S_t) + \gamma \, \mathrm{Skew}(S_t) + \varepsilon_t \tag{5.6}$$

where $F_{t,T} - S_T$ is the *ex-post* futures premium, $F_{t,T}$ is the futures price in t for delivery in T, S_T is the spot price in T, $\mathrm{Var}(S_t)$ is the variance of daily spot prices at time t, and $\mathrm{Skew}(S_t)$ is the skewness of daily spot prices at time t.

Note that, due to the effect of banking restrictions on Phase I spot prices (see [1]), we only consider Phase II spot prices from BlueNext in our empirical application. Thus, the data covers the period going from February 26, 2008 to March 31, 2009 (i.e. a sample of 282 daily observations). Futures prices are taken from ECX to compute the futures premium variable. To replicate the results, the data for this chapter

Table 5.1 R Code: Estimating Bessembinder and Lemmon's (2002) Structural Model

```
1  path<-"C:/"
2  setwd(path)
3  library(stats)
4  data=read.csv("data_chapter5.csv",sep=",")
5  attach(data)
6  futpremium=data[,2]
7  varspot=data[,3]
8  skewspot=data[,4]
9  modelfutpremium<-lm(futpremium~varspot+skewspot)
10 summary(modelfutpremium)
11 layout(matrix(1:4,2,2))
12 plot(modelfutpremium)
13 acf(residuals(modelfutpremium))
14 acf(residuals(modelfutpremium),type='partial')
```

Table 5.2 R Output: Estimating Bessembinder and Lemmon's (2002) Structural Model

```
Call:
lm(formula = futpremium ~ varspot + skewspot)

Residuals:
     Min       1Q    Median       3Q      Max
-1.47396  -0.31160  -0.01327  0.24460  1.86381

Coefficients:
            Estimate Std. Error t value Pr(>|t|)
(Intercept) -0.02630    0.05086  -0.517    0.606
varspot     -0.02185    0.01352  -1.617    0.107
skewspot     0.01554    0.06298   0.247    0.805

Residual standard error: 0.4523 on 279 degrees of freedom
Multiple R-squared: 0.01311,   Adjusted R-squared: 0.006033
F-statistic: 1.853 on 2 and 279 DF,  p-value: 0.1587
```

is available at: http://sites.google.com/site/jpchevallier/publications/books/springer/data_chapter5.csv.

Equation (5.6) can be estimated based on the R code given in Table 5.1.

Lines 1 to 8 load the relevant library and data. Line 9 performs the estimation, along with parameter estimates and diagnostic checks given in lines 10 to 14. The output is reproduced in Table 5.2.

As predicted by Bessembinder and Lemmon's (2002, [7]) model, we observe that the futures premium for CO_2 allowances is negatively related to the variance of the spot price, and positively related to the unstandardized skewness of the spot price. However, only the α coefficient for the variance of the spot price can be identified

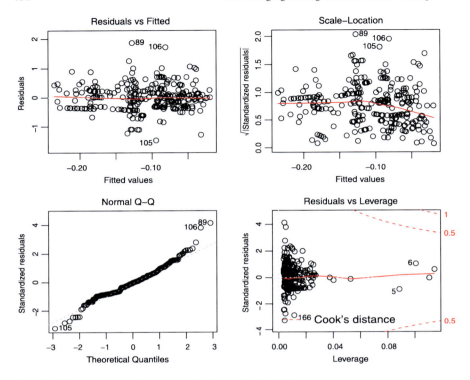

Fig. 5.2 Diagnostic plots for the model estimated in Eq. (5.6)

as being statistically significant (at the 10% level). Therefore, we cannot conclude as in [18] that the skewness of spot prices plays a crucial role for the risk assessment of market participants. This latter result may be due to the somewhat shorter data sample in our illustrative example.

Besides, we may conclude from the visual examination of Figs. 5.2 and 5.3 that the residuals of the model are not auto-correlated.

Overall, our findings support Bessembinder and Lemmon's (2002, [7]) model applied to the EU ETS. Indeed, the virtual inexistence of storage costs for CO_2 allowances leads us to consider another class of models compared to the theory of storage. The model developed in this section allows to derive interesting properties in terms of variance and skewness of spot prices. These results globally indicate a negative relationship between ex-post risk premia and the variance of CO_2 spot prices. We have therefore achieved a better understanding of the relationships between CO_2 allowances spot and futures prices.[9] Incorporating risk premia yields indeed to making informed hedging decisions in the banking and finance industries,

---

[9]Note that a cointegration exercise between CO_2 allowances spot and futures prices, in the spirit of the exercises studied in Chap. 4, can be found in Chevallier (2010, [19]).

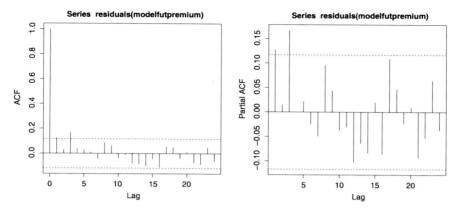

Fig. 5.3 Autocorrelation and partial-autocorrelation functions for the residuals from the model estimated in Eq. (5.6)

while allowing regulated utilities to better relate futures prices to movements of spot prices.

Next, we examine how carbon price risks intervene in the power producers' fuel-switching behavior, with an application to the case of the UK.

5.3 Managing Carbon Price Risk in the Power Sector

The switching from coal to gas represents the lowest abatement cost expected in power generation. Besides, power generators need to take into account carbon costs since the creation in January 2005 of the EU ETS. The EU trading scheme provides a carbon price signal that helps internalizing the external costs of greenhouse gas emissions, especially in power generation, which weights more than 50% of allowances distributed.

This section investigates how power operators' fuel-switching decisions are affected by the relative change in fuel and carbon prices under the EU ETS, and to what extent emission reductions may be achieved at lowest cost with an application to the UK.

The results highlight a negative (positive) relationship between coal (gas) price changes and the decision to switch in the energy mix of power producers in the UK, after taking into account carbon costs in the range of 15–20 Euro/ton of CO_2 during 2005–2008. These results are carefully identified following a decomposition in full- and sub-samples (2005–2006 and 2007–2008) in order to account for time-varying fuel prices.

5.3.1 Economic Rationale

We examine the impact of fuel (gas, coal) and carbon prices on UK electricity generators' behaviors under the EU ETS. We aim at understanding how fuel-switching decisions are impacted by (i) the relative variation of fuel prices as an input to power producers' energy mix in the UK, and (ii) the introduction of carbon costs in Europe since 2005. Indeed, electricity generation from coal is cheaper than from natural gas. However, coal is nearly twice as carbon-intensive as gas, and the introduction of carbon costs may make gas more interesting than coal to produce electricity. Thus, fuel-switching decisions in the UK may have been greatly impacted by the CO_2 price path.

Previous literature identifies fuel-switching as an important channel of energy shocks transmission to the economy. Concerning the EU ETS, Voorspools and D'haeseleer (2006, [42]) study the impact of the CO_2 price on cross-border electricity trade through the modelling of eight interconnected European zones. Delarue and D'haeseleer (2007, [22]) estimate a potential of 9.5% emissions reduction by fuel-switching in the Belgian-based electrical power system during the summer. Delarue et al. (2008, [23]) show evidence of fuel-switching during the summer 2005 both in Germany (through a cross-border switch with the Netherlands) and in the UK (through internal switch). Besides, Ellerman and Feilhauer (2008, [26]) demonstrate that fuel-switching took place in Germany during the summer 2005 when gas was at its cheapest price and load sufficiently low. Finally, McGuinness and Ellerman (2008, [35]) use an econometric model of fuel-switching to explain the utilization rate of a plant with demand, fuel prices, the carbon price and the coal-to-gas ratio as explanatory variables. Despite the overall increase in coal generation in the UK (because of high gas prices), they conclude that the increase in demand would have been larger without the introduction of the EU ETS.

The results presented in this section provide policy guidance concerning the impact of fuel and carbon price changes on fuel switching decisions in the UK power sector. Indeed, we find a negative (positive) relationship between coal (gas) price changes and the decision to switch in the energy mix of power producers in the UK, after taking into account carbon costs in the range of 15–20 Euro/ton of CO_2 during 2005–2008. These results are carefully identified following a decomposition in full- and sub-samples (2005–2006 and 2007–2008) in order to account for time-varying fuel prices, and the respective peaks in the time-series of coal and gas spot prices used by electricity generators in the UK.

Before proceeding to the actual econometric analysis, let us start by examining the main characteristics of the power sector in the UK.

5.3.2 UK Power Sector

Power generation contributes to more than 50% of emissions reductions in the EU ETS [3, 4]. The main reasons are (i) the perceived ability of this sector to reduce emissions with low abatement costs (through the switching between coal and gas or the possibility to use low-carbon technologies such as renewable energy or carbon capture and storage), and (ii) the non-exposure of this sector to international, non-EU competition in opposition to the main industrial sector (cement, steel, glass

Fig. 5.4 Coal, natural gas and nuclear use in power generation in the UK (1970–2008). Source: UK Department for Energy and Climate Change (DECC)

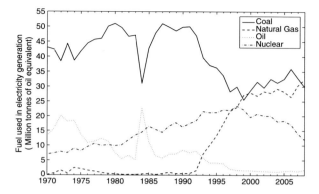

production, etc.). We choose to focus on the UK electricity market which presents an interesting energy mix.

As shown in Fig. 5.4, the important share of coal and gas in the electricity mix in the UK (39% of coal and 36% of gas) makes possible the opportunity to switch from coal to gas depending on the price of the carbon price in order to minimize the electricity production cost. The opportunity to switch provides an interesting solution for power generators to reduce their emissions at low cost. In order to benefit from these cheap abatement opportunities, electricity generators need to compare the CO_2 price and marginal abatement costs. It may be cheaper to switch for a less carbon-intensive fuel (such as gas) instead of using coal and buying permits.

Next, we detail the main factors influencing the fuel-switching behaviour of power producers.[10]

5.3.3 Factors Influencing Fuel-Switching

Several factors impact the opportunity for fuel-switching, such as the influence of fuel and CO_2 prices. First, concerning fuel prices, the switching point may be seen as the carbon price at which unused available gas-fired capacity would be substituted for coal-fired generation (see more details in Chap. 2). For any given couple of coal and gas-fired plants connected to the electricity grid, there is a single switching point. However, as there are many pairs of plants with different efficiencies and fuel prices, it leads to a switching band. The lower bound of the band corresponds to the substitution of the most efficient and lowest cost unused gas-fired plant with the least efficient and highest cost coal-fired plant in service. The upper bound is the substitution of the most efficient and lowest cost coal plant by the least efficient highest cost gas plant (Ellerman and Feilhauer (2008, [26])).

---

[10]The interested reader may refer to Chap. 2 for a full description of the fuel-switching mechanism in the power sector, in presence of carbon costs.

Second, the variation of the carbon price constitutes another important (and relatively new) risk factor. Delarue et al. (2008, [23]) compare the potential for emissions reduction in Europe at different levels of carbon prices. At 20 Euro/ton of CO_2, there exists a significant reduction potential during the summer. This is consistent with Sijm et al. [40, 41], Chen et al. (2008, [14]) and Lise et al. (2010, [31]), who estimated the switching price of CO_2 around 18.5 Euro/ton of CO_2. The amount of 60 Euro/ton of CO_2 allows a significant reduction throughout the year, while 120 Euro/ton of CO_2 leads to a constant reduction during the whole year. Finally, around 150 Euro/ton of CO_2, all the switching opportunities are used, and no further abatement may be reached by using fuel switching only.

Third, the potential for fuel-switching is very dependent on the load. Indeed, at full load, during winter peak-hours for instance, all the plants are running so that no opportunity to switch exists. The switch can only occur if coal-plants are running while some gas-fired plants are available to replace them. The best opportunities occur when the load is relatively low and mostly met by coal-fired-plants, i.e. during weekends, nights and summers. Then, with adequate economic incentives, available gas-fired units may be switched. Consequently, the possibility of fuel-switching varies throughout the year depending on the season (winter or summer), the time of the week (day-of-week or week-end effects), and the period of the day (day or night).

As highlighted by Delarue and D'haeseleer (2007, [22]) and Delarue et al. (2008, [23]), the gas-to-coal ratio is the main variable of interest in the EU ETS. Given the large share of allowances distributed to the electricity sector (50% on average), power producers thus constitute the main players on the carbon market [3, 4]. Therefore, their ability to switch between their fuel inputs (e.g. coal or gas) to produce one MWh of electricity is key to understanding the amount of abatement that may be achieved within the scheme. This explains why we do not consider other variables of interest, such as oil. Moreover, oil in the UK is used only to meet high peak demand especially in winter and its fuel share is very low (1%), so it is unlikely to interfere with coal and gas switching. Besides, gas and coal are the cheapest alternatives, while oil has encountered a high level of volatility over the period.

In the next section, we develop the econometric analysis.

5.3.4 Econometric Analysis

We present first the data used, and second the econometric specification.[11]

5.3.4.1 Data

The natural gas price used is the National Balancing Point (NBP) spot price (see Fig. 5.5). The NBP is a pricing and delivery point for natural gas in the UK. It is

---

[11]The data used for this section is not available for download.

Fig. 5.5 NBP gas daily spot price (2005–2008). Source: Reuters

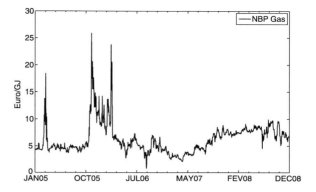

Fig. 5.6 ARA coal daily spot price (2005–2008). Source: Reuters

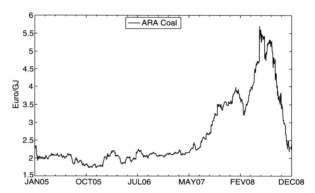

the most liquid gas trading market in Europe, and has a major influence on the price that European consumers pay for their gas. As such, the NBP price represents the best proxy of the European gas market price determined close to end-users. The NBP price is expressed in pence per therm. It is converted in Euro per GJ using the conversion factors[12] from the International Energy Agency, and the daily exchange rate for EUR-GBP from the European Central Bank (ECB).

The coal price is the Antwerp—Rotterdam—Amsterdam (ARA) coal spot price (see Fig. 5.6), which is the major imported coal in northwest Europe. The ARA coal is expressed in US$ per ton. It is converted into Euro/GJ using the calorific value[13] from the UK Department for Business, Enterprise and Regulatory Reform,[14] and the daily exchange rate USD-EUR from the ECB.

Interestingly, we may observe asymmetric patterns in the price changes of gas and coal prices: the time-series of the NBP gas spot price exhibits a strong increase during the period going from September 2005 to January 2006, while the time-series of the ARA coal price has a distinct peak during the period going from April

---

[12] 1 therm = 0.10550559 GJ.

[13] The coal calorific value is equal to 24.9 MJ/kg.

[14] Now, it is called the UK Department for Business, Innovation and Skills.

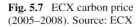

Fig. 5.7 ECX carbon price (2005–2008). Source: ECX

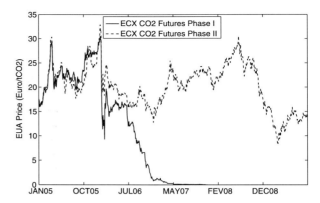

to November 2008. These distinctive features of fuel prices variation may have an important effect on power producers' fuel-switching behaviour in the UK. This intuition will be the basis for the estimation of our econometric model in full vs. sub-samples in order to disentangle the likely influence of time-varying fuel prices.

The carbon price is the December 2008 futures price for European Union Allowances (EUAs) valid for trading under the EU ETS (see Fig. 5.7). It is traded in Euro/ton of CO_2 on ECX. Since Phase I carbon spot prices did not send a reliable price signal (Alberola and Chevallier (2009, [1])), we choose to work with the December 2008 futures contract instead.

The methodology to compute the switch price has been detailed in Chap. 2. We simply recall here that the switch variable represents the threshold for the carbon price above which it becomes profitable for an electric power producer to switch from coal to natural gas, and below which it is beneficial to switch from natural gas to coal. Therefore, the switch price may be seen as a shadow price. It represents the theoretical equilibrium between the clean dark and clean spark spreads (thereby explicitly taking into account CO_2 costs) that is being used by power producers to produce one unit of electricity out of gas- or coal-fired plants.[15] As long as the carbon price is below this switching price, coal plants are more profitable than gas plants—even after taking carbon costs into account. This switching price is more sensitive to natural gas price changes than to coal prices changes (Kanen (2006, [29])).

The UK fuel-switching price is computed according to this methodology by plugging in the values of 35% for the net thermal efficiency of a conventional coal-fired plant, 0.95 ton of CO_2/MWh for the CO_2 emissions factor of a conventional coal-fired power plant, 49.13% for the net thermal efficiency of a conventional gas-fired

---

[15]Note that we not include in the analysis the capital and operations and maintenance costs of coal vs. natural gas which may result in the unusually low spreads noted in Fig. 5.8. Hence we aim at capturing the short-term effect of introducing emissions trading on power producers' fuel-switching behaviour (see Delarue et al. (2010, [24])). A longer term analysis of fuel substitution with additional costs associated with a retro-fit is not possible in this setting.

Fig. 5.8 UK daily
fuel-switching price
(2005–2008). Source:
calculation from the author
by using Reuters, ECX

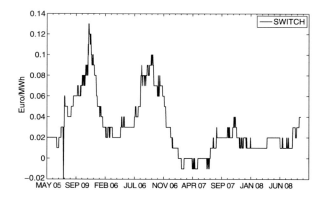

Table 5.3 Descriptive statistics

	SWITCH	NBP	ARA	EUA
Mean	0.0303	41.8075	91.5798	21.1473
Median	0.0200	35.7000	68.7900	21.1850
Maximum	0.1300	168.0000	224.3000	32.2500
Minimum	−0.0200	5.0000	50.7800	12.2500
Std. Dev.	0.0273	19.9449	43.9318	3.5406
Skewness	1.0655	1.6154	1.3087	0.1776
Kurtosis	3.6132	7.8649	3.5075	2.8525
Observations	872	872	872	872

Note: SWITCH refers to the switch price, NBP to the gas spot price, ARA to the coal spot price, and EUA to the carbon price

plant, 0.41 ton of CO_2/MWh for the CO_2 emissions factor of a conventional gas-fired power plant, and by correcting for CO_2 costs (see Fig. 5.8).

The descriptive statistics for all time series used in this section are shown in Table 5.3. Next, we present the econometric model used.

5.3.4.2 Econometric Model

In order to test for the main risk factors linked to the impact of energy prices and emissions trading on fuel-switching in the UK power sector, we run the following regression:

$$switch_t = \alpha + \psi NBP_t + \phi ARA_t + \gamma EUA_t + \varepsilon_t \qquad (5.7)$$

where $switch_t$ is the switching price between coal and gas in order to produce one MWh of electricity as described in Chap. 2, corrected for carbon costs. α is the constant term, NBP_t is the UK National Balance Point gas price, ARA_t is the coal price, EUA_t is the carbon price, and ε_t is the error term.

We also test for possible ARCH effects (Engle and Bollerslev (1986, [27])) by using the standard GARCH(1, 1) specification in the variance equation:

$$\sigma_t^2 = \omega + \alpha\varepsilon_{t-1}^2 + \beta\sigma_{t-1}^2 \tag{5.8}$$

The estimation is carried out by QML with the Bollerslev-Wooldridge procedure to generate robust standard errors. After applying the usual unit root tests (ADF, KPSS, PP) on all price series, we may conclude that they are integrated of order 1 ($I(1)$). Thus, all price series are made stationary by using the first-difference logarithmic transformation (which is convenient for interpretation in terms of percentage change). Stationarity test results for the main variable of interest, $switch_t$, are shown in Table 5.4.

In Eq. (4), various lags have been tested for the dependent variable ($switch_t$), as well as for the independent variables (NBP_t, ARA_t, EUA_t) by following the Box-Jenkins methodology of lags identification for ARMA(p, q) processes. No lags were found to be statistically significant at usual confidence levels, and thus they have been removed from reduced-form estimates.

We comment in the next section the estimation results.

5.3.5 Empirical Results

In this section, we comment the results obtained during the full period (i.e. by running Eqs. (5.7) and (5.8) on the full sample), as well as in sub-samples (i.e. by dividing the full sample in two equal sub-samples and running Eqs. (5.7) and (5.8) again) in order to account for time-varying coal and gas price changes in the UK during 2005–2008.

5.3.5.1 Full Sample Results

Estimation results of Eqs. (5.7) and (5.8) are presented in Table 5.5. The quality of the regressions is verified by the following diagnostic tests: the Adjusted R-squared, the Akaike Information (AIC), the Schwarz information criterion (SC), the Ljung-Box residuals autorocorrelation test (LB Test), the Engle test for ARCH effects (ARCH test), and the p-value of the F-Statistic (F-Stat). These tests are validated in all regressions. Besides, we verify that the residuals are not autocorrelated (LB test > 5%). The presence of ARCH effects is taken into account through the GARCH(1, 1) specification in the variance equation (where all coefficients are significant and positive).

In Table 5.5, while the constant term is highly significant and negative, we observe most importantly that the sign of the NBP gas and ARA coal coefficients are negative: high variations of fuel prices tend to affect negatively the fuel-switching price (when the switch price becomes negative, it is cheaper to burn gas than coal). However, we also notice that the coefficient for the NBP gas price is not statistically

Table 5.4   Unit root tests for the *switch_t* variable

Null Hypothesis: D(SWITCH) has a unit root

Exogenous: None

Lag Length: 0 (Automatic based on SIC, MAXLAG = 20)

	t-Statistic	Prob.*
Augmented Dickey-Fuller test statistic	−36.3062	0.0000
Test critical values:		
1% level	−2.5676	
5% level	−1.9411	
10% level	−1.6164	

*MacKinnon (1996, [34]) one-sided p-values

Null Hypothesis: D(SWITCH) has a unit root

Exogenous: None

Bandwidth: 10 (Newey-West using Bartlett kernel)

	Adj. t-Stat	Prob.*
Phillips-Perron test statistic	−37.6033	0.0000
Test critical values:		
1% level	−2.5676	
5% level	−1.9411	
10% level	−1.6164	

*MacKinnon (1996, [34]) one-sided p-values

Null Hypothesis: D(SWITCH) is stationary

Exogenous: Constant

Bandwidth: 12 (Newey-West using Bartlett kernel)

	LM-Stat.
Kwiatkowski-Phillips-Schmidt-Shin test statistic	0.0783
Asymptotic critical values*:	
1% level	0.7390
5% level	0.4630
10% level	0.3470

*Kwiatkowski-Phillips-Schmidt-Shin (1992, [30])

Note: ADF and PP: Model 1 (without intercept or trend) has been chosen with a lag truncation of 0 based on the Schwarz Information Criterion (SIC). As the value calculated for the t-statistic is strictly inferior to the critical value at 1% confidence level, we reject the null hypothesis of nonstationarity. KPSS: Model 2 without trend has been chosen with a lag truncation of 12 according to the Newey-West lag detection method based on Bartlett kernels. We accept the null hypothesis of stationarity, as the LM statistic calculated is inferior to the critical value at the 1% confidence level

Table 5.5 Estimation results of Eqs. (5.7) and (5.8) during the full-sample

	Coefficient t	Std. Error	Prob.
Mean Equation			
α	−0.0118	0.0008	0.0001
ψ	−0.0001	0.0001	0.1513
ϕ	−0.0001	0.0001	0.0001
γ	0.0020	0.0001	0.0001
Variance Equation			
ω	0.0001	0.0001	0.0001
α	0.8790	0.1060	0.0001
β	0.2565	0.0730	0.0004
Diagnostic tests			
Adjusted R-squared	0.2200		
AIC	−6.0669		
SC	−6.0286		
Log likelihood	2652.1848		
LB Test	0.1457		
ARCH Test	0.3243		
F-Stat.	0.0001		

Note: Std. Error is the standard error, Prob. is the statistical significance probability, AIC is the Akaike Information, SC the Schwarz information criterion, LB Test is the Ljung-Box test, ARCH test is the Engle ARCH test, and F-Stat. the p-value of the F-Statistic

significant. The central economic intuition behind this result is that changes in fuel-switching of power operators in the UK are driven mainly by coal price changes (which have a statistically significant impact on the switch price at the 1% level), and not by natural gas price changes (which are not a significant explanatory variable in this regression).

Our intuitions are confirmed by re-estimating the econometric model without the NBP daily gas spot price. In Table 5.6, we notice first the remarkable stability of the coefficients estimated, which constitutes a robustness check of the validity of our econometric specification. Second, the statistically significant and negative effect of the coal price on the switch price remains valid in the reduced form econometric model without the non-significant NBP gas price. Third, we observe that the coefficient for the carbon price is statistically significant and positive, which means that at higher levels of carbon prices, it becomes profitable to switch from coal (the most carbon-intensive fuel in terms of emissions) to gas (which is less carbon-intensive). The latter result also applies for Table 5.5.

In the variance equations of Tables 5.5 and 5.6, the measure of persistence $(\alpha + \beta)$ is close to the value of one, which means that the variance process is close to being integrated (Engle and Bollerslev (1986, [27])). In the next section, we test whether the same conclusions hold when re-estimating the econometric model of Eqs. (5.7) and (5.8) in sub-samples in order to account for time-varying energy price changes.

Table 5.6 Estimation results of Eqs. (5.7) and (5.8) without NBP during the full-sample

	Coefficient t	Std. Error	Prob.
Mean Equation			
α	−0.0111	0.0008	0.0001
ϕ	−0.0001	0.0001	0.0001
γ	0.0019	0.0001	0.0001
Variance Equation			
ω	0.0001	0.0001	0.0001
α	0.8747	0.1091	0.0001
β	0.2599	0.0755	0.0006
Diagnostic tests			
Adjusted R-squared	0.2109		
AIC	−6.0645		
SC	−6.0316		
Log likelihood	2650.1269		
LB Test	0.1578		
ARCH Test	0.3146		
F-Stat.	0.0001		

Note: Std. Error is the standard error, Prob. is the statistical significance probability, AIC is the Akaike Information, SC the Schwarz information criterion, LB Test is the Ljung-Box test, ARCH test is the Engle ARCH test, and F-Stat. the p-value of the F-Statistic

5.3.5.2 Sub-Samples Results

Estimation results for sub-samples #1 (2005–2006) and #2 (2007–2008) may be found in Tables 5.7 and 5.8, respectively.

The results differ starkly from the full-sample results, and call for the following comments. First, the coefficient for the ARA coal price remains negative and statistically significant in all regressions. This confirms our main result, i.e that the switch price is mainly driven by coal price changes (be it in full or sub-periods).

Second, we obtain opposite signs between the full- and sub-periods estimates for the gas coefficient. Besides, the gas coefficient is not significant in the full-sample, while it is significant in both sub-samples at the 1% level. The main reason behind these counter-intuitive results lies in the peak in the price of gas (as seen in Fig. 5.5), which may bias the results obtained during the full period.[16] Thus, the rationale behind fuel-switching (as detailed in Chap. 2) remains valid in the UK power sector when taking into account both carbon costs and the peak in the time-series of the NBP natural daily spot price. In sub-samples, the gas price has a statistically significant effect on power producers' decisions when deciding whether they should produce one unit of electricity out of coal-fired or gas-fired plants.

This section allows us to re-establish this main economic message including carbon price risks: power producers do arbitrate between their fuel inputs in order to

---

[16]See Chap. 2 for structural break tests.

Table 5.7 Estimation results of Eqs. (5.7) and (5.8) during the sub-sample #1

	Coefficient t	Std. Error	Prob.
Mean Equation			
α	0.2866	0.0053	0.0001
ψ	0.0002	0.0001	0.0001
ϕ	−0.0032	0.0001	0.0001
γ	−0.0027	0.0001	0.0001
Variance Equation			
ω	0.0001	0.0001	0.0001
α	0.9590	0.0764	0.0001
β	0.0390	0.0050	0.0043
Diagnostic tests			
Adjusted R-squared	0.2198		
AIC	−5.6739		
SC	−5.6083		
Log likelihood	1241.0754		
LB Test	0.1490		
ARCH Test	0.5029		
F-Stat.	0.0001		

Note: Std. Error is the standard error, Prob. is the statistical significance probability, AIC is the Akaike Information, SC the Schwarz information criterion, LB Test is the Ljung-Box test, ARCH test is the Engle ARCH test, and F-Stat. the p-value of the F-Statistic

Table 5.8 Estimation results of Eqs. (5.7) and (5.8) during the sub-sample #2

	Coefficient t	Std. Error	Prob.
Mean Equation			
α	−0.0248	0.0053	0.0001
ψ	0.0004	0.0001	0.0001
ϕ	−0.0001	0.0001	0.0001
γ	0.0012	0.0001	0.0001
Variance Equation			
ω	0.0001	0.0001	0.0001
α	0.5676	0.1201	0.0001
β	0.4257	0.0880	0.0001
Diagnostic tests			
Adjusted R-squared	0.2837		
AIC	−7.3486		
SC	−7.2832		
Log likelihood	1612.6787		
LB Test	0.1809		
ARCH Test	0.7774		
F-Stat.	0.0001		

Note: Std. Error is the standard error, Prob. is the statistical significance probability, AIC is the Akaike Information, SC the Schwarz information criterion, LB Test is the Ljung-Box test, ARCH test is the Engle ARCH test, and F-Stat. the p-value of the F-Statistic

produce electricity (after accounting for carbon costs) depending on relative fuel-prices. Thus, the decomposition in sub-samples allows us to uncover the critical role played by time-varying energy prices (for gas and coal mainly) in the fuel-switching behavior of power producers in the UK under the EU ETS.

Finally, we observe that the sign of the coefficient for the carbon price is negative in sub-sample #1 and positive in sub-sample #2, while it is positive during the full-sample. We may conclude from these different regressions that carbon costs are obviously a very important input (as the EUA_t variable is statistically significant in all regressions) for power producers when deciding on their energy mix, and that the sign of this relationship should be positive: higher costs for CO_2 emissions translate into higher costs for the final users of electricity according to the cost pass-through which is a salient characteristic of electricity markets.[17]

5.3.6 Summary

Thanks to its suitable fuel mix (39% of coal and 36% of gas), the UK is reported to have the greatest potential for emission reduction through fuel switching within the EU. The switching from coal to gas represents the lowest abatement cost expected in power generation. It has several advantages because it is technologically feasible to implement, leads to a more efficient use of gas plants (through CCGT for instance)[18] and allows significant emissions reduction.

This section documents how to manage carbon price risks in the power sector, with an application to the UK. We find that the switch price is impacted by fuel and carbon prices during 2005–2008. When accounting for a carbon price in the range of 15–20 Euro/ton of CO_2, not only does the ARA coal spot price negatively affects the potential for fuel-switching (at higher levels of coal prices, it becomes profitable to switch to gas-fired power plants), but also it becomes necessary to take into account time-varying fuel prices in order to correctly determine this empirical relationship. Indeed, the full-sample results do not reveal any statistically significant effect of the NBP daily gas spot price on power producers' fuel-switching behaviour in the UK. A positive relationship between gas price changes and the switch price may be re-established in sub-samples decomposition, by explicitly taking into account the wide variation of gas price changes during the first sub-sample (2005–2006), at a

---

[17]Note that the negative sign found in subsample #1 for the coefficient of the carbon price may be explained by confusing price signals sent to power producers during 2005–2007, as the carbon spot price geared towards zero due to the presence of banking restrictions between 2007 and 2008 (Alberola and Chevallier (2009, [1])).

[18]Demand is negatively linked with Combined Cycle Gas Turbines (CCGT), and positively with coal. This is consistent with the fact that increasing demand would yield to a decrease in the number of gas-fired plants and in the fuel-switching opportunity. If the carbon price increases, the use of CCGT is encouraged and the use of coal is lower. The decrease in the use of coal is met partly with the use of gas, but it also encourages the use of other low-carbon technologies (such as nuclear or renewable energy).

time where carbon prices were following an erratic price path too (see Alberola et al. (2008, [2])).

For policy makers, these results yield insights into how the introduction of carbon costs in the power sector is transformed into opportunities for emissions reduction through fuel-switching. This section also helps to understand how electricity generators are able to trade-off between carbon prices and emission reduction costs.

In the next section, we consider other risk-hedging strategies specific to CO_2 allowances from the perspective of portfolio management.

5.4 Portfolio Management

In this section, we investigate the properties of carbon assets for portfolio management. In previous chapters, we have established that the preferences of the marginal investor differ between the stock, bond and commodity markets on the one hand, and the carbon market on the other hand. These results are relevant for portfolio management: when markets are segmented, risk prices differ. From that perspective, carbon allowances may be used for diversification purposes.

First, we present the asset management strategies with energy, weather, bond and equity variables. Second, we detail the optimal composition for such a globally diversified portfolio based on mean-variance optimization.

5.4.1 Composition of the Portfolio

Let us consider a database[19] going from April 2005 to January 2009, and composed of the following assets:

- Carbon prices: EUAs and CERs from the European Climate Exchange (ECX);
- Energy prices: Henry-Hub Natural Gas Futures, Coal Futures Month Ahead Forward Cost Insurance Freight from the Amsterdam—Rotterdam—Antwerp region (CIF ARA), Zeebrugge Natural Gas Next Month Forward, Off-Peak Electricity Next Month Forward from the European Energy Exchange (EEX), Electricity Powernext Baseload Next Month Forward, Coal Rotterdam Futures, London Brent Crude Oil Futures and Crude Oil Futures from the New York Mercantile Exchange (NYMEX);
- Weather derivatives: Weather Futures and Climate Futures Eco Clean Energy Index from the Intercontinental Exchange (ICE);
- Bonds: Five-Year Euro Benchmark Bond from the European Central Bank (ECB), One-Month US Treasury Bill (T-Bill), Bond Schatz and Bond Bulb Treasury Bills;

---

[19]This data is not reproduced for this chapter.

- Equities: Euronext 100 Price Index, Dow Jones EuroSTOXX 50 Price Index, Standard and Poor's Euro Price Index.

The rationale behind the inclusion of energy variables along the more traditional equity and bond investments is that carbon, oil, gas and coal can have a diversifying effect in asset management [15]. Besides, weather derivatives allow to hedge the risks attached to temperatures changes, and thus to increases or decreases in CO_2 emissions, as detailed in Chap. 2. References [10] and [6] recall that diversification can reduce risk substantially. Hence, the main logic behind composing such a portfolio not only with bonds and equities, but also with energy commodities, is to achieve a lower level of risk because its individual asset components do not always move together. Next, we assess the expected return that can be achieved in a stylized exercise based on mean-variance optimization.

Box 5.1 (Measuring Portfolio Performance with the Sharpe Ratio) The Sharpe ratio (Sharpe (1966, [39])) measures the ratio of return from a portfolio to volatility. This ratio may be seen as a risk-adjusted measure of return that is often used to evaluate the performance of a portfolio. It is defined as:

$$S = \frac{R_t - R_t^f}{\sigma_t} \tag{5.9}$$

with R_t the asset return, R_t^f the risk-free rate (such as the T-Bill rate), and σ_t the standard deviation of the excess of the asset return. Therefore, the Sharpe ratio is simply the risk premium per unit of risk. Given its computation, the higher the Sharpe ratio, the better the return for each unit of risk. Under the Capital Asset Pricing Model (CAPM, [33, 38]), the portfolio on the efficient frontier with the highest Sharpe ratio is the market portfolio.

5.4.2 Mean-Variance Optimization and the Portfolio Frontier

The database detailed above contains carbon assets (EUAs and CERs), energy assets (oil, coal, natural gas), weather derivatives, bonds and equities. We discuss here how to implement asset management strategies.

The portfolio frontier is a set of portfolios with a given expected return. Following [10] and [6], we wish to pick the portfolio with the minimum variance. This choice embodies trade-offs between risk and return, and represents the level of risk aversion for a specific group of investors.

Among all portfolios that have a given expected return $E(\cdot)$, the optimization problem consists in choosing the portfolio with the minimum variance $V(\cdot)$:

$$V(R) = \sum_{n=1}^{N} w_n^2 V(R_n) + 2 \sum_{n<m} w_n w_m \, \text{Cov}(R_n, R_m) \tag{5.10}$$

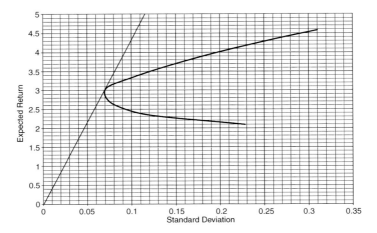

Fig. 5.9 Portfolio frontier analysis

$$1 - \sum_{n=1}^{N} w_n \tag{5.11}$$

$$E(R) = \sum_{n=1}^{N} w_n E(R_n) + \left(1 - \sum_{n=1}^{N} w_n\right) R_f = E \tag{5.12}$$

with N the total number of risky assets, R_n the return on the risky asset, R_f the return on the riskless asset, R_m the return on the market portfolio, $w_n, n = \{1, \ldots, N\}$ the portfolio weights of the risky assets to minimize. Equation (5.11) gives the weight of the riskless asset (i.e. US T-Bills). Assuming that we only care about mean and variance [6, 10], we only need to consider portfolios on the portfolio frontier.

This problem can be solved in Excel (by using the solver, see [10]), or in R (by using Rmetrics,[20] see [44]). Formally, the portfolio frontier is delimited by the line linking the riskless asset with the tangent portfolio.[21] That is why, in Fig. 5.9, we draw the line linking that portfolio to the riskless asset.

By considering the line linking the riskless asset with the points on the hyperbola, we notice in Fig. 5.9 that the optimal portfolio composed of energy (including carbon), weather, bond, equity risky assets and the US T-Bill riskless asset achieves a standard deviation equal to 0.07 for an expected return of 3%. Thus, for investors as a group, the demand will be a combination of the tangent portfolio and the riskless asset. This result allows to minimize the variance for a given level of expected return, and illustrates the benefits of diversification by adding energy and weather variables to more traditional equity and bond assets. Indeed, we are able to verify

---

[20] See https://www.rmetrics.org/.

[21] At the market equilibrium, demand equals supply and the tangent portfolio coincides with the market portfolio.

the standard property that diversification outside of a group of assets allows to reduce idiosyncratic risk. However, it cannot eliminate systematic risk. Note that very risk-averse agents would choose a portfolio closer to the riskless asset, while less risk-averse agents would choose a portfolio even above the tangent portfolio. Note also that this solution corresponds also to the highest Sharpe ratio for that portfolio.

Collectively, these results illustrate the benefits of introducing carbon assets for diversification purposes in portfolio management (see [15]). As discussed in Chap. 2, risk factors on the carbon market are mainly linked to institutional decision changes by the EU Commission and to the power producers' fuel-switching behavior. Finally, we shall keep in mind that there is a dependency of all types of assets to macroeconomic shocks, as shown by the recent 'credit crunch' financial crisis.

Appendix: Computing Implied Volatility from Option Prices

In the financial economics literature, risk-hedging strategies may also be designed by resorting to option prices (see [20] for an application to EUAs by using a dataset of plain vanilla European option prices at-the-money). We briefly recall here how to derive implied volatility measures from option pricing.

Let $C(\tau, K)_{obs}$ be the observed call option price, and $C(\tau, K, \sigma)_{BS}$ the Black-Scholes (BS) price computed using the implied volatility σ. By definition, we have $C(\tau, K)_{obs} = C(\tau, K, \sigma)_{BS}$. The implied volatility of the strike price is obtained by numerically inverting the BS formula, which can be done by solving:

$$\min_{\sigma}\left(C(\tau, K)_{obs} - C(\tau, K, \sigma)_{BS}\right)^2 \qquad (5.13)$$

Table 5.9 Descriptive statistics for CO_2 prices

	EUA	DIFFEUA
Mean	16.3413	−0.0087
Median	14.7400	0.0001
Maximum	28.7300	1.2200
Minimum	7.9600	−1.6600
Std. Dev.	4.6158	0.3838
Skewness	1.0476	−0.5157
Kurtosis	2.8900	4.9257
Jarque-Bera	130.7879	141.5801
Probability	0.0001	0.0001
Observations	713	712
Diagnostic tests		
Ljung-Box test DIFFEUA		0.8470
Ljung-Box test Squared DIFFEUA		0.0030
Engle ARCH test		0.0034

Note: The Box-Pierce test is computed with 20 lags. The Engle ARCH test is computed with one lag

Reference [20] shows that option prices on the carbon market lead to pricing errors that are usual for commodity or equity markets (see also [5]). Implied volatility series can then be used to represent the risk-neutral probability distribution of a given asset in financial economics.

Problems

Problem 5.1 (Calibrating GARCH Models for Carbon Prices)

(a) Consider the descriptive statistics for the CO_2 price series in Table 5.9: how well do they seem to fit the GARCH(p, q) models specifications?

Table 5.10 GARCH estimates for CO_2 prices

Dependent Variable: DIFFEUA
Method: ML—ARCH (BHHH)—Normal distribution
Sample (adjusted): 2/28/2008 12/09/2010
Included observations: 711 after adjustments
Bollerslev-Wooldridge robust standard errors & covariance

	Coefficient	Std. Error	z-Statistic	Prob.
C	−0.0008	0.0113	−0.0780	0.9378
DIFFEUA(-1)	0.0220	0.0392	0.5615	0.5745
	Variance Equation			
C	0.0004	0.0006	0.6677	0.5043
RESID(-1)^2	0.0567	0.0166	3.4105	0.0006
GARCH(-1)	0.9407	0.0176	53.2424	0.0001
R-squared	0.0018	Mean dependent var		−0.0087
Adjusted R-squared	−0.0038	S.D. dependent var		0.3840
S.E. of regression	0.3848	Akaike info criterion		0.7395
Sum squared resid	104.5431	Schwarz criterion		0.7717
Log likelihood	−257.9242	F-statistic		0.3266
Durbin-Watson stat	1.9135	Prob(F-statistic)		0.8601
Diagnostic tests				
Ljung-Box test DIFFEUA		0.8060		
Ljung-Box test Squared DIFFEUA		0.8630		
Engle ARCH test		0.8629		

Note: The Box-Pierce test is computed with 20 lags. The Engle ARCH test is computed with one lag

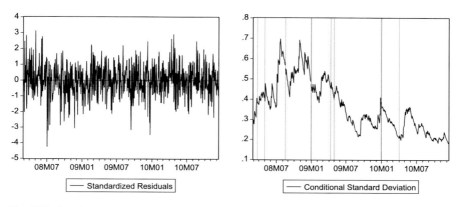

Fig. 5.10 Standardized innovations (*left panel*) and conditional standard deviations (*right panel*) of the GARCH(p, q) model

Table 5.11 EGARCH estimates for CO_2 prices

Dependent Variable: DIFFEUA
Method: ML—ARCH (BHHH)—Student's t distribution
Sample (adjusted): 2/28/2008 12/09/2010
Included observations: 711 after adjustments

	Coefficient	Std. Error	z-Statistic	Prob.
C	0.0028	0.0114	0.2491	0.8032
DIFFEUA(-1)	0.0066	0.0387	0.1709	0.8643
Variance Equation				
C(3)	-0.1361	0.0353	-3.8555	0.0001
C(4)	0.1494	0.0350	4.2585	0.0001
C(5)	0.0022	0.0205	0.1073	0.9145
C(6)	0.9909	0.0076	128.9354	0.0001
R-squared	-0.0001	Mean dependent var		-0.0087
Adjusted R-squared	-0.0072	S.D. dependent var		0.3840
S.E. of regression	0.3854	Akaike info criterion		0.7320
Sum squared resid	104.7504	Schwarz criterion		0.7705
Log likelihood	-254.2318	Durbin-Watson stat		1.8854
Diagnostic tests				
Ljung-Box test DIFFEUA		0.4840		
Ljung-Box test Squared DIFFEUA		0.9963		
Engle ARCH test		0.9935		

Note: The Box-Pierce test is computed with 20 lags. The Engle ARCH test is computed with one lag

Table 5.12 TGARCH estimates for CO_2 prices

Dependent Variable: DIFFEUA
Method: ML—ARCH (BHHH)—GED
Sample (adjusted): 2/28/2008 12/09/2010
Included observations: 711 after adjustments

	Coefficient	Std. Error	z-Statistic	Prob.
C	0.0030	0.0113	0.2677	0.7889
DIFFEUA(-1)	0.0024	0.0386	0.0622	0.9504
Variance Equation				
C	0.0002	0.0005	0.4520	0.6512
RESID(-1)^2	0.0631	0.0210	3.0034	0.0027
RESID(-1)$^2 *$ (RESID(-1) < 0)	-0.0157	0.0210	-0.7489	0.4539
GARCH(-1)	0.9440	0.0158	59.6354	0.0001
GED PARAMETER	1.5396	0.1196	12.8659	0.0001
R-squared	-0.0006	Mean dependent var		-0.0087
Adjusted R-squared	-0.0091	S.D. dependent var		0.3840
S.E. of regression	0.3858	Akaike info criterion		0.7269
Sum squared resid	104.8051	Schwarz criterion		0.7719
Log likelihood	-251.4432	Durbin-Watson stat		1.8779
Diagnostic tests				
Ljung-Box test DIFFEUA		0.4730		
Ljung-Box test Squared DIFFEUA		0.8940		
Engle ARCH test		0.8951		

Note: The Box-Pierce test is computed with 20 lags. The Engle ARCH test is computed with one lag

(b) Following the Box-Jenkins methodology, we configure the ARMA(p, q) processes that provide the best fit to the CO_2 time-series. Then, we estimate the corresponding GARCH(p, q) model for the CO_2 price series:

$$Y_t = \theta X_t' + \varepsilon_t \tag{5.14}$$

$$\sigma_t^2 = \omega + \sum_{i=1}^{p} \alpha_i \varepsilon_{t-i}^2 + \sum_{j=1}^{q} \beta_j \sigma_{t-j}^2 \tag{5.15}$$

with σ_t^2 the conditional variance, which is function of a constant term ω, the ARCH term ε_{t-i}^2, and the GARCH term σ_{t-j}^2.

In Table 5.10, comment on:

- the statistical significance of the parameters obtained in the mean and variance equations;

- the value of the Ljung-Box test;
- the value of the ARCH test.

In Fig. 5.10, comment on:

- the plot of the standardized innovations of the GARCH(p, q) model;
- the plot of the conditional standard deviations of the GARCH(p, q) model.

(c) We consider other models than the standard GARCH(p, q) model with a Gaussian conditional probability distribution. First, we use the specification for the conditional variance proposed by Nelson (1991, [36]):

$$\log(\sigma_t^2) = \omega + \sum_{i=1}^{p} \alpha_i \left| \frac{\varepsilon_{t-i}}{\sigma_{t-i}} \right| + \sum_{j=1}^{q} \beta_j \log(\sigma_{t-j}^2) + \sum_{k=1}^{r} \gamma_k \frac{\varepsilon_{t-k}}{\sigma_{t-k}} \tag{5.16}$$

where γ tests for the presence of the leverage effect. Recall that the leverage effect implies a higher level of volatility associated to decreasing prices in the financial economics literature. The EGARCH model is estimated with a Student t distribution.

Second, we use the asymmetric TGARCH model by Zakoian (1994, [45]):

$$\sigma_t^2 = \omega + \sum_{i=1}^{p} \alpha_i \varepsilon_{t-i}^2 + \sum_{j=1}^{q} \beta_j \sigma_{t-j}^2 + \sum_{k=1}^{r} \gamma_k \varepsilon_{t-k}^2 \Gamma_{t-k} \tag{5.17}$$

where $\Gamma_t = 1$ if $\varepsilon_t < 0$, and 0 otherwise. $\varepsilon_{t-i} > 0$ and $\varepsilon_{t-i} < 0$ denote, respectively, good and bad news. The TGARCH model is estimated with a Generalized Error Distribution (GED).

Consider the Tables 5.11 and 5.12: do you observe any improvement with the alternative specification structures in the variance equation?

References

1. Alberola E, Chevallier J (2009) European carbon prices and banking restrictions: evidence from Phase I (2005–2007). Energy J 30:51–80
2. Alberola E, Chevallier J, Cheze B (2008) Price drivers and structural breaks in European carbon prices 2005-07. Energy Policy 36:787–797
3. Alberola E, Chevallier J, Cheze B (2008) The EU emissions trading scheme: the effects of industrial production and CO_2 emissions on European carbon prices. Int Econ 116:93–125
4. Alberola E, Chevallier J, Cheze B (2009) Emissions compliances and carbon prices under the EU ETS: a country specific analysis of industrial sectors. J Policy Model 31:446–462
5. Barone-Adesi G, Engle RF, Mancini L (2008) A GARCH option pricing model with filtered historical simulation. Rev Financ Stud 21:1223–1258
6. Berk J, DeMarzo P (2008) Corporate finance. Pearson International Edition, New York
7. Bessembinder H Lemmon ML (2002) Equilibrium pricing and optimal hedging in electricity forward markets. J Finance 57:1347–1382
8. Blyth W, Bunn D, Kettunen J, Wilson T (2009) Policy interactions, risk and price formation in carbon markets. Energy Policy 37:5192–5207
9. Blyth W, Bunn D (2011) Coevolution of policy, market and technical price risks in the EU ETS. Energy Policy 39:4578–4593

10. Bodie Z, Kane A, Marcus AJ (2008) Investments, 7th edn. McGraw Hill, New York
11. Breeden DT (1980) Consumption risks in futures markets. J Finance 35:503–520
12. Brennan M (1958) The supply of storage. Am Econ Rev 48:50–72
13. Broadie M, Chernov M, Johannes M (2007) Model specification and risk premia: evidence from futures options. J Finance 62:1453–1490
14. Chen Y, Sijm J, Hobbs BF, Lise W (2008) Implications of CO_2 emissions trading for short-run electricity market outcomes in northwest Europe. J Regul Econ 34:251–281
15. Chevallier J (2009) Energy risk management with carbon assets. Int J Glob Energy Issues 32:328–349
16. Chevallier J (2009) Carbon futures and macroeconomic risk factors: a view from the EU ETS. Energy Econ 31:614–625
17. Chevallier J (2009) Modelling the convenience yield in carbon prices using daily and realized measures. Int Rev Appl Financ Issues Econ 1:56–73
18. Chevallier J (2010) Modelling risk premia in CO_2 allowances spot and futures prices. Econ Model 27:717–729
19. Chevallier J (2010) A note on cointegrating and vector autoregressive relationships between CO_2 allowances spot and futures prices. Econ Bull 30:1564–1584
20. Chevallier J, Ielpo F, Mercier L (2009) Risk aversion and institutional information disclosure on the European carbon market: a case-study of the 2006 compliance event. Energy Policy 37:15–28
21. Cootner PH (1960) Returns to speculators: Telser vs Keynes. J Polit Econ 68:396–404
22. Delarue E, D'haeseleer WD (2007) Price determination of ETS allowances through the switching level of coal and gas in the power sector. Int J Energy Res 31:1001–1013
23. Delarue E, Voorspools KR, D'haeseleer WD (2008) Fuel switching in the electricity sector under the EU ETS: review and prospective. J Sol Energy Eng 134:40–58
24. Delarue E, Ellerman AD, D'haeseleer WD (2010) Short-term CO_2 abatement in the European power sector: 2005–2006. Clim Change Econ 1:113–133
25. Dusak K (1973) Futures trading and investor returns: an investigation of commodity market risk premiums. J Polit Econ 81:1387–1406
26. Ellerman AD, Feilhauer S (2008) A top-down and bottom-up look at emissions abatement in Germany in response to the EU ETS. Working paper 2008-17, Centre for Energy and Environmental Policy Research, MIT, USA
27. Engle RF, Bollerslev T (1986) Modelling the persistence of conditional variances. Econom Rev 5:1–50
28. Kaldor N (1939) Speculation and economic stability. Rev Econ Stud 7:1–27
29. Kanen JLM (2006) Carbon trading and pricing. Environmental Finance Publications, London
30. Kwiatkowski DP, Phillips PCB, Schmidt P, Shin Y (1992) Testing the null hypothesis of stationarity against the alternative of the unit root: how sure are we that economic time series are non stationary? J Econom 54:159–178
31. Lise W, Sijm J, Hobbs BF (2010) The impact of the EU ETS on prices, profits and emissions in the power sector: simulation results with the COMPETES EU20 model. Environ Resour Econ 47:23–44
32. Longstaff FA, Wang A (2004) Electricity forward prices: a high-frequency empirical analysis. J Finance 59:1877–1900
33. Markowitz H (1959) Portfolio selection: efficient diversification of investments. Wiley, New York
34. MacKinnon JG (1996) Numerical distribution functions for unit root and cointegration tests. J Appl Econom 11:601–618
35. McGuinness M, Ellerman AD (2008) CO_2 abatement in the UK power sector: evidence from the EU ETS trial period. Working paper 2008-04, Centre for Energy and Environmental Policy Research, MIT, USA
36. Nelson DB (1991) Conditional heteroskedasticity in asset returns: a new approach. Econometrica 59:347–370
37. Pindyck R (2001) The dynamics of commodity spot and futures markets: a primer. Energy J 22:1–29

38. Sharpe WF (1964) Capital asset prices: a theory of market equilibrium under conditions of risk. J Finance 19:425–442
39. Sharpe WF (1966) Mutual fund performance. J Business 39:119–138
40. Sijm JPM, Bakker SJA, Chen Y, Harmsen HW, Lise W (2006) CO_2 price dynamics: a follow-up analysis of the implications of EU emissions trading for the price of electricity. Working paper 06-015, Energy Research Centre for the Netherlands, the Netherlands
41. Sijm JPM, Hers SJ, Lise W, Wetzelaer BJHW (2008) The impact of the EU ETS on electricity prices. Working paper 08-007, Energy Research Centre for the Netherlands, the Netherlands
42. Voorspools KR, D'haeseleer WD (2006) Modelling of electricity generation of large interconnected power systems: how can a CO_2 tax influence the European generation mix. Energy Convers Manag 47:1338–1358
43. Working H (1949) The theory of the price of storage. Am Econ Rev 39:1254–1262
44. Würtz D, Chalabi Y, Chen W, Ellis A (2009) Portfolio optimization with R/Rmetrics. Rmetrics Publishing, Zurich
45. Zakoian JM (1994) Threshold heteroskedastic models. J Econ Dyn Control 18:931–955

Chapter 6
Advanced Topics: Time-to-Maturity and Modeling the Volatility of Carbon Prices

Abstract This chapter examines the maturity effects of the CO_2 futures contracts traded on the European Climate Exchange, in conjunction with the CO_2 spot prices traded on the BlueNext market. It investigates the 'Samuelson hypothesis' that the volatility of futures price changes increases as a contract's delivery date nears. While the literature on commodities usually finds strong empirical support of this hypothesis (in agricultural markets for instance), this chapter provides a weak support for the CO_2 market. Volatility is found to increase near the expiration of the contract only with realized volatility measures (constructed as the sum of intraday squared returns). The net cost-of-carry and GARCH modeling approaches fail to detect such time-to-maturity effects. This chapter illustrates the superiority of realized volatility in carbon pricing, as the data is observed and modeled at the highest possible frequency. An Appendix completes this chapter by dealing with statistical tests to detect instability in the volatility of carbon prices.

6.1 The Relationship Between Volatility and Time-to-Maturity in Carbon Prices

Volatility is central for forecasting, hedging, and risk-management purposes. It may be measured from daily data—separating the conditional and unconditional components of volatility by using GARCH models—or from intraday data—which consists in computing 'realized volatility' as the sum of intraday squared returns. In this chapter, we use both measures of volatility to document the behavior of the CO_2 futures prices with respect to time-to-maturity.

Samuelson (1965, [37]) argued that the futures price volatility increases as the futures contract approaches its expiration date. This prediction is sufficiently well-known that it is widely referred to as the 'Samuelson hypothesis'. This relationship is important to margin setting [13] and for hedging strategies. Depending on whether this relation is positive or negative, hedgers should choose futures contracts with either a short or long time-to-maturity, so that the price volatility is minimized. Traders might consider switching to contracts further away from expiration day. Otherwise, they will face higher volatility and require a higher risk premium [24]. In markets where the Samuelson hypothesis holds, the accurate valuation of options or futures-related derivatives requires that estimates of futures volatility depend on the time remaining until the underlying futures contract matures [11].

Previous literature features empirical studies of the Samuelson hypothesis for several commodity markets. Adrangi and Chatrath (2003, [1]) investigate nonlinear dynamics in the futures prices of coffee, sugar and cocoa. Their study confirms the Samuelson hypothesis for these futures price series. McMillan and Speight (2004, [30]) examine time-to-maturity effects in FTSE-100 index futures based on intra-day and GARCH modeling. On equity markets, their results contradict increasing volatility near time-to-maturity for futures contracts, other than for the very high frequency data. On US gas markets, Movassagh and Modjtahedi (2005, [31]) and Mu (2007, [32]) identify larger standard deviations in the means of futures contracts as the month to maturity nears. Their results thus confirm that natural gas futures volatility declines with contract horizon. Haff et al. (2008, [25]) also find support for the Samuelson effect in the UK natural gas market. Kalev and Duong (2008, [26]) test the Samuelson hypothesis in commodity futures markets using 'realized range'.[1] They find strong support for agricultural futures, unlike metal and financial futures. Duong and Kalev (2008, [24]) confirm this result by using daily GARCH models and intraday realized volatility models. Besides, they use the 'negative co-variance' hypothesis[2] to harness the power of their tests.

The CO_2 futures market raises interesting research questions: is the Samuelson effect present within CO_2 futures? What are the implications of the hypothesis for this new market? What does the presence of this effect say about market partici-pants? Can we expect differences between this market and more established mar-kets? To answer these questions, this chapter documents the volatility behavior of carbon prices with respect to time-to-maturity. Using daily and intraday data, we measure the Samuelson effect by an increase in the standard deviation of the futures contract price changes as the time-to-maturity approaches.

The econometric analysis unfolds in three steps. First, by using daily data for CO_2 spot and futures prices, we follow Bessembinder et al. (1995, [10]) to com-pute the net carry cost, and instrument it to determine the relationship with time-to-maturity. Second, based on GARCH modeling with daily data from April 2005 to December 2009, we test for the presence of maturity effects in the returns of the CO_2 futures prices, by adding the time-to-maturity component in the variance equa-tion. Third, we use intraday data to compute the realized volatility of the December 2008 and December 2009 futures contracts traded on ECX, and regress it against the time-to-maturity, spot prices volatility, seasonality and liquidity variables. The dataset contains two years of tick-by-tick data from ECX CO_2 emissions futures, corresponding to the 2008 and 2009 futures contracts.[3]

---

[1] Martens and van Dijk (2007, [29]) define realized range as the sum of all high-low range for each intraday interval.

[2] The negative covariance hypothesis states that the Samuelson hypothesis is more likely to be supported in markets that exhibit a negative covariance between spot price changes and changes in net carry costs [10].

[3] The choice to start our analysis with the 2008 contract is motivated by (i) the erratic behavior of *spot* prices during 2005–2007 due to banking restrictions (Alberola and Chevallier 2009, [2]),

This estimation strategy yields the following results: while the net carry cost approach and GARCH modeling do not allow to detect the presence of a Samuelson effect in the CO_2 futures prices, the use of high-frequency data clearly shows that the realized volatility of the CO_2 futures prices increases near the expiration date. Thus, the higher the frequency of the data observed, the higher the accuracy of volatility measures when dealing with carbon assets. The superiority of realized *vs.* daily volatility measures in estimation has been shown in previous literature on equity markets (see Andersen, Bollerslev, Diebold, and Labys (henceforth ABDL (2001, 2003), [3, 4], and [30]). Therefore, our results bring new insights into the empirical characteristics of CO_2 prices in terms of volatility, and thus may be of use to brokers, market operators and risk managers alike to incorporate directly these features in futures pricing.

Next, we provide more background information on the Samuelson hypothesis.

6.2 Background on the Samuelson Hypothesis

Samuelson (1965) states that volatility is a function of time-to-maturity, and that volatility increases as a futures contract approaches its maturity. The intuition is that most of the relevant information is revealed when the contract is nearing maturity. Early in a futures contract's life, little information is known about the future spot price for the underlying commodity. Later, as the contract nears maturity, the rate of information acquisition increases (Kolb 1991, [27]). Consequently, the futures price of the commodity does not necessarily respond instantly and significantly to new information coming into the market. As the futures contract approaches maturity, the price must converge to the spot price, and therefore it tends to react more strongly to the arrival of new information. This relationship also gives a hint whether the market is under- or over-reacting to the information flow.

The rationale for investigating the temporal pattern of futures price volatility may be explained in many aspects. According to Board and Sutcliff (1990, [13]), the relationship between time-to-maturity and volatility is important for margin setting. The desired margin is set on the basis of the futures price volatility, and may be seen as a positive function of price volatility. Therefore, if the futures price volatility increases as the futures contract approaches maturity, the cash balances held by traders to cover for margin calls should also be increased as the maturity date approaches (Duong and Kalev 2008, [24]).

Another reason lies in the fact that volatility is a key factor of interest for hedgers and speculators. On the one hand, hedgers adjust the hedge ratio based on the variability in the futures price (so as to offset their position in the spot market). They choose between futures contracts with short or long time-to-maturity in order to

---

which were less robust than *futures* for price signaling in the medium-term; and (ii) the validity of the 2008 contract during Phase II (2008–2012), which offers the 'bankability' of CO_2 emissions allowances until the end of Phase III (2013–2020).

minimize the price volatility. On the other hand, speculators are interested in price volatility to benefit from arbitrage opportunities (and thereby provide liquidity to the market).

Interestingly, Anderson and Danthine (1983, [5]) give us another explanation of why the volatility-maturity relationship in futures contracts is important. They argue that it is not the time-to-expiration which determines the volatility, but the degree to which uncertainty is resolved depending on the information flow. According to this view, more uncertainty tends to be resolved (or more information flows into the market) as delivery approaches. If most of the information arrives into the market near the maturity period, then it leads to a greater level of volatility and confirms the Samuelson hypothesis. However, if the information flow resolves the uncertainty at earlier stages of the contract, then volatility will increase also at earlier stages of the life-cycle of the futures contracts. These arguments appear especially important to understand the presence (or not) of the Samuelson hypothesis in the carbon futures market, which is associated with a high level of regulatory uncertainty (Alberola and Chevallier 2009, [2]).

Contrary to this view, Bessembinder et al. (1996, [11]) argue that while it is certainly possible that a systematic clustering of information flows near futures delivery dates could cause commensurate increases in volatility at those times, there is nonetheless no compelling explanation why information flows should cluster near futures maturity dates. In the CO_2 market, allowances are not only traded near maturity dates, but throughout the year with strong price changes during the months of February to May of each year (the so-called 'compliance events', see Chevallier et al. 2009, [18]).

Last but not least, the relationship between volatility and maturity needs to be taken into account when pricing options on futures. The accurate valuation of options or related derivatives on futures indeed requires knowledge of the dynamic evolution of futures price volatility [11]. Next, we present the data used.

6.3 Data

Various exchanges offer spot and futures trading of CO_2 allowances in the EU ETS. We focus on CO_2 spot and futures prices exchanged on the most liquid platforms: BlueNext (spot), and ECX (futures). In what follows, we detail the data used with both daily and intraday frequencies throughout the chapter.[4]

6.3.1 Daily Frequency

Spot and futures price series are shown in Fig. 6.1. From October 2006 until December 2007, CO_2 spot prices have been decreasing towards zero due to the banking

---

[4]Note that the data used in this chapter is not available for download.

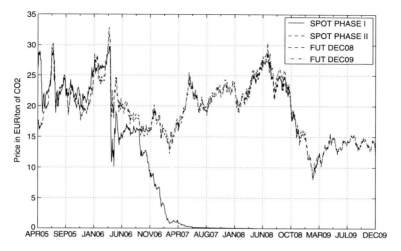

Fig. 6.1 BlueNext CO_2 spot prices valid for Phase I and Phase II and ECX EUA December 2008 and December 2009 futures contracts from April 22, 2005 to December 15, 2009. Source: BlueNext, European Climate Exchange (ECX)

restrictions implemented between 2007 and 2008 (Alberola and Chevallier (2009, [2])). Due to this erratic and non-reliable behavior of spot prices during Phase I, we choose to work only with Phase II CO_2 spot prices in this chapter. The trading of CO_2 spot prices valid for Phase II started on BlueNext on February 26, 2008. Thus, the start of the second trading period of the EU ETS corresponds to the start of the dataset for spot prices. Our sample of Phase II CO_2 spot prices contains 479 observations for the years 2008 and 2009, which will be used for the determination of the net carry cost.

Since Phase II futures contracts conveyed a coherent price signal during 2005–2007 (around 20 Euro/ton of CO_2 as shown in Fig. 6.1), the study period for CO_2 futures prices goes from April 22, 2005 until December 15, 2009, i.e. the date at which the December 2009 contract expires. Our sample of December 2008 (2009) futures prices contains 937 (1,208) observations, which appears sufficient for the GARCH estimation.

Descriptive statistics for all dependent variables used in the chapter may be found in Table 6.1. The distributional properties of CO_2 spot and futures prices appear non-normal.

6.3.2 Intraday Frequency

Our sample contains two years of tick-by-tick transactions for the ECX futures contracts of maturity December 2008 and December 2009, going from January 2 to December 15 for each year. This is equivalent to 240 (244) days of trading after cleaning the data for outliers, and until the expiration of the 2008 (2009) contract. The

Table 6.1 Descriptive statistics

Variable	Mean	Median	Max	Min	Std. Dev.	Skew.	Kurt.	N
Spot Phase II	17.3313	14.7600	28.7300	7.9600	5.3470	0.4967	1.8039	479
Futures 2008	20.9317	20.9000	32.2500	12.2500	3.5784	0.2068	2.7840	937
Futures 2009	19.6061	19.9500	32.9000	8.2000	4.7083	0.0342	2.3479	1208
RV 2008	0.0152	0.0064	0.3293	0.0001	0.0307	6.4926	57.5542	240
RV 2009	1.2603	0.8801	5.4715	0.0428	1.0282	1.8899	7.0781	244

Note: *Spot PhaseII* refers to BlueNext spot prices valid during Phase II (2008–2012) of the EU ETS, *Futures* 2008 to the ECX December 2008 futures contract, *Futures* 2009 to the ECX December 2009 futures contract, *RV* 2008 to the realized volatility measure of the ECX December 2008 futures contract, and *RV* 2009 to the realized volatility measure of the ECX December 2009 futures contract. *Std.Dev.* stands for Standard Deviation, *Skew.* for Skewness, *Kurt.* for Kurtosis, and *N* for the number of observations

total amount of intraday observations in our sample is equal to 167,004 (370,718) for the December 2008 (2009) contract. The average amount of transactions for the ECX CO_2 emissions futures tick-by-tick data is equal to 700 (1,500) trades per day for the December 2008 (2009) contract. This corresponds to an average of 60 (30) seconds between each transaction for the December 2008 (December 2009) contract. Intraday data is used for the measure of realized volatility. This level of transactions appears comparable to the values found for other markets. For instance, Thomakos and Wang (2003, [38]) note that the average number of price changes per day is 163 for the Eurodollar, 3,366 for the S& P500, and 1,710 for T-bonds.

The use of only two years of data is mainly justified by the lack of high-frequency data availability, since the EU ETS was created in 2005 and exchanges emerged incrementally to record transactions on this market. Since the end of 2007, both the liquidity of the EU ETS and the availability of high-frequency data have been increasing. Besides, the choice of our sample size is also motivated by the non-reliable price signal for CO_2 allowances valid during Phase I (2005–2007) of the EU ETS. This argument is due to the non-bankability of allowances from 2007 to 2008, which led to the disconnection between Phase I and II prices.[5]

Next, we start our investigation of the Samuelson hypothesis in CO_2 futures prices by following the net carry cost approach.

6.4 The 'Net Carry Cost' Approach

The cost of carry relation is a no-arbitrage condition dating from Working (1949, [39]) and Brennan (1958, [14]), which links futures prices to contemporaneous spot

---

[5]Alberola and Chevallier (2009, [2]) indeed notice a divergence between Phase I spot and futures prices—which decreased towards zero—and Phase II futures prices—which conveyed a medium-term price signal around 20 Euro/ton of CO_2 throughout the historical data available for the second phase of the scheme.

prices by the per period cost to the marginal trader of holding the underlying asset in inventory. Bessembinder et al. (1996, [11]) link negative covariation between spot prices and the cost of carrying inventory changes to mean reversion in spot prices, by considering relations between shocks to the current spot price and the revision in expectations regarding the future spot price. To induce the holding of inventory when prices are mean-reverting, the predictable capital loss (gain) in the wake of a positive (negative) price shock must be offset by changes in proportional carrying costs if markets are to clear.

To summarize, the Samuelson hypothesis requires either systematic increases in spot return volatility near each futures expiration date (which is highly implausible given that futures contracts mature throughout the year), or negative covariation between spot returns and the cost of carrying inventory. These conditions are most likely to be met in markets for real assets, especially those where convenience yields display substantial intertemporal variation. The most realistic explanation for substantial time variation in inventory carrying costs derives from the variation of real service flows (i.e. the 'convenience yield'). Analogous arguments for financial assets (and CO_2 allowances) cannot easily be made, since it is difficult to postulate either substantial variation in asset inventory or the existence of convenience yields.

According to Bessembinder et al. (1995, [10]), the Samuelson hypothesis is more likely to be supported in markets that exhibit a negative covariance between spot price changes and changes in net carry costs, including agricultural markets, crude oil, and to a lesser extent metals. We detail this approach below.

6.4.1 Computational Steps

We compute the net carry cost, defined by Bessembinder et al. (1995, [10]) as being the difference between the natural logarithms of the December 2008 (December 2009) contract and the CO_2 spot price, weighted by the time-to-maturity[6] remaining until the expiration of the futures contract as specified below:

$$c_t = \frac{(f_t - s_t)}{TTM_t} \qquad (6.1)$$

where c_t is the net carry cost, f_t the natural logarithm of the ECX December 2008 (December 2009) futures price, s_t the natural logarithm of CO_2 spot prices, and TTM_t the time-to-maturity (in days) of the December 2008 (2009) futures contract. Given the non-reliable behavior of spot prices during Phase I, we consider only Phase II CO_2 spot and futures prices.

The graphs of the net carry cost (NCC) for the December 2008 and December 2009 futures contract are shown in Fig. 6.2. We observe a sharp increase in volatility as the time-to-maturity decreases. This behavior seems to validate visually the

---

[6]For a financial asset noted $A(t, T)$ with t the current period and T the terminal period, the time-to-maturity may be written as $A(t, T) = A(\tau)$ with $\tau = T - t$.

Fig. 6.2 Net carry cost
(NCC) for the December
2008 (*top panel*) and
December 2009 (*bottom
panel*) carbon futures
contracts from February 26,
2008 to December 15, 2009.
Source: BlueNext, European
Climate Exchange (ECX)

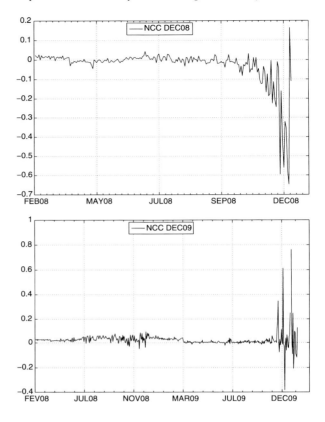

Samuelson hypothesis.[7] We develop in the next section a formal regression analysis
to test statistically this effect.

6.4.2 Regression Analysis

The covariance between changes in spot prices and changes in net carry costs is
estimated as follows (Duong and Kalev (2008, [24])):

$$\Delta c_t = \alpha_0 + \alpha_1 \Delta S_t + \varepsilon_t \qquad (6.2)$$

where Δc_t is the change in net carry costs for the ECX December 2008 (December
2009) futures contract, ΔS_t the change in spot prices at time t, and ε_t the error
term. For the negative covariance hypothesis to hold, α_1 needs to be statistically
significant and *negative*. To facilitate the comparison of coefficient estimates, we
divide each price series by its own mean (Bessembinder et al. 1996, [11]).

---

[7]Note a similar visual inspection for the ECX December 2007 futures contract may be found in
Chevallier (2009, [16]).

Table 6.2 Testing the 'negative covariance' hypothesis in the CO_2 allowance market using the net carry cost approach

Note: Newey-West robust standard errors are presented in parentheses below coefficient estimates. ***, **, *indicate statistical significance at, respectively, the 1%, 5% and 10% levels. Δc_t is the change in net carry costs for the ECX December 2008 (*DEC*08) and December 2009 (*DEC*09) futures contracts. The model estimated is described in Eq. (6.2). The regressions are estimated in first differences, and the price series are scaled by dividing them with their own means

Δc_t	DEC08 (1)	DEC09 (2)
α_0	-0.0496^{**}	-0.0198^{*}
	(0.0202)	(0.0116)
α_1	-0.6794^{*}	-0.4204
	(0.3887)	(0.7387)
Diagnostic tests		
R-squared	0.0132	0.0015
F-Statistic	0.1048	0.3887
Durbin-Watson	1.1360	2.1206
Ljung-Box	0.7910	0.1803
White test	0.9806	0.3622
AIC	0.6093	0.2054
SC	0.5764	0.2286
N	205	478

6.4.3 Empirical Results

Regression results are shown in Table 6.2. The results are obtained with the Newey-West heteroskedasticity consistent covariance matrix procedure. The following diagnostic tests and statistics are given: the adjusted R-squared, the p-value of the F-test statistic (*F-Statistic*), the Durbin-Watson statistic (*Durbin-Watson*), the p-value of the Ljung-Box portmanteau test for serial correlation with 20 lags (*Ljung-Box*), the p-value of the White heteroskedasticity test (*White test*), the Akaike Information Criterion (*AIC*) and the Schwarz Criterion (*SC*).

Similarly to Duong and Kalev (2008, [24]) for agricultural futures, we first find statistical support of the 'negative covariance' hypothesis for the ECX December 2008 futures contract. In regression (1), α_1 is *negative* and statistically significant at the 10% level. However, this result is not robust for the ECX December 2009 futures contract. In regression (2), α_1 is negative but not significant. Thus, we may cautiously conclude from these first results that the covariance between changes in spot prices and changes in net carry costs is insignificant in explaining the Samuelson hypothesis in the CO_2 allowance market.[8] This may be due to the specificities of carbon allowances as a new commodity without the usual storage costs.[9]

---

[8]Note the low levels of the adjusted R-squared obtained in both regressions are consistent with the values found by Duong and Kalev (2008, [24]) for energy metals and financial futures.

[9]Indeed, CO_2 allowances have a virtual existence on the balance sheets of regulated companies, and are therefore free of storage costs (unlike barrels of oil or other storable energy commodities). See Chap. 5 for more details.

Besides, there may be a possibility that the net carry cost approach based on the linear regression model is not suitable to identify statistically the Samuelson maturity effect. We further explore this possibility in the next section based on GARCH modeling.

6.5 GARCH Modeling

This section investigates whether the Samuelson hypothesis applies to the CO_2 futures prices by using daily data in a GARCH modeling framework. The basic insight of the econometric strategy consists in testing whether the time-to-maturity has a statistically significant effect on the returns of the CO_2 futures prices, by including it as an exogenous regressor in the variance equation. This econometric framework has been adopted, among others, by Chen et al. (1999, [15]) for the Nikkei Index futures and Adrangi and Chatrath (2003, [1]) for coffee, sugar and cocoa futures.

6.5.1 GARCH Specification

Following the Box-Jenkins procedure to investigate the autocorrelation structure of the ECX December 2008 and December 2009 futures returns, we specify an autoregressive process of order 1. This approach is also in line with previous literature for CO_2 price modeling using GARCH models (Benz and Trueck (2009, [9]), Oberndorfer 2009, [34]). Thus, the AR(1)-GARCH(1, 1) model[10] estimated is:

$$r_t = \mu + \rho r_{t-1} + \varepsilon_t, \quad \varepsilon_t \mid \Omega_{t-1} \sim i.i.d.(0, \sigma_t^2) \qquad (6.3)$$

$$\sigma_t^2 = \omega + \alpha \varepsilon_{t-1}^2 + \beta \sigma_{t-1}^2 + \delta TTM_t \qquad (6.4)$$

where r_t is the futures contract log-return for day t, r_{t-1} is the AR(1), and ε_t is the error term process with mean zero and conditional variance σ_t^2. μ and ω are constant variables in, respectively, the mean and variance equations. ε_{t-1}^2 and σ_{t-1}^2 are, respectively, the ARCH and GARCH terms in the variance equation. TTM_t is the number of days to maturity.[11]

The AR(1)-GARCH(1, 1) model is estimated by Quasi Maximum Likelihood (QML) using Bollerslev-Wooldridge robust standard errors and the BHHH algorithm (Berndt et al. 1974, [12]). Innovations follow a Student's t-distribution to accommodate the leptokurtic characteristic of the daily CO_2 time-series (Paoella and Taschini 2008, [35], Daskalakis et al. 2009, [23]).

---

[10] As in Adrangi and Chatrath (2003, [1]), the estimation of other specifications such as the Exponential GARCH(1, 1) or the Asymmetric component GARCH(1, 1) leads to the same conclusions.

[11] As in Bessembinder et al. (1996, [11]), we have experimented various regressions with $log(TTM_t)$ or $\sqrt{TTM_t}$ without altering qualitatively the results obtained. This comment applies in the remainder of the chapter.

Table 6.3 Testing the Samuelson hypothesis in the CO_2 allowance market using the AR(1)-GARCH(1, 1) model

r_t	DEC08 (3)	DEC09 (4)
μ	0.0013^{**}	0.0011^{*}
	(0.0007)	(0.0006)
ρ	0.1389^{***}	0.0843^{***}
	(0.0328)	(0.0299)
ω	0.0001^{**}	0.0001^{***}
	(0.0001)	(0.0001)
α	0.1364^{***}	0.1704^{***}
	(0.0361)	(0.0367)
β	0.7943^{***}	0.7838^{***}
	(0.0503)	(0.0415)
δ	-0.0001	-0.0001
	(0.0001)	(0.0001)
Diagnostic tests		
R-squared	0.0118	0.0099
Log-Likelihood	2157.3600	2776.373
Ljung-Box	0.6280	0.1890
Engle ARCH	0.7487	0.8600
N	936	1207

Note: Bollerslev-Wooldridge robust standard errors are presented in parentheses below coefficient estimates. ***, **, * indicate statistical significance at, respectively, the 1%, 5% and 10% levels. r_t is the log-returns for the ECX December 2008 (*DEC*08) and December 2009 (*DEC*09) futures contracts. The model estimated is described in Eq. (3)

6.5.2 Empirical Results

Regression results are provided in Table 6.3. As diagnostic tests, we provide for each regression the Ljung-Box portmanteau test for serial correlation, and the Engle ARCH test. The AR(1), as well as the ARCH and GARCH coefficients, are positive and statistically significant in explaining the log-returns of ECX December 2008 and December 2009 futures contracts. Clearly, the standard setting of GARCH models allows to capture the heteroskedatic behavior of CO_2 futures returns highlighted in Chap. 3. This result is in line with Benz and Trueck (2009, [9]) and Oberndorfer (2009, [34]).

However, the Samuelson hypothesis is not evident in the CO_2 allowance market based on GARCH modeling, as the δ coefficient for the time-to-maturity variable is negative but not statistically significant for both futures contracts (regressions (3) and (4)). These results are consistent with the net carry cost approach, and tend to confirm that the Samuelson hypothesis is not supported in the CO_2 allowance market. We explore in the next section whether high-frequency data might lead to a different diagnostic.

6.6 Realized Volatility Modeling

This last section investigates whether the Samuelson hypothesis applies to the CO_2 allowance market by using intraday data. The econometric strategy consists in measuring 'realized' volatility from high frequency data, and then to regress this measure of volatility against the time-to-maturity. As shown in previous literature on equity markets [3, 4, 30], this high-frequency measure of volatility may indeed reveal different results compared to the net carry cost approach and standard GARCH estimates based on daily data.

6.6.1 Computational Steps

Let $p(t)$ be the logarithmic asset price at time t. Assume that the returns are sampled on a Δ-period yielding $r_{t,\Delta} \equiv p(t) - p(t - \Delta)$. The *realized variance* is defined as the sum of the corresponding $1/\Delta$, which is assumed to be an integer for simplicity, high-frequency intraday squared returns, or:

$$RV_{t+1}(\Delta) \equiv \sum_{j-1}^{1/\Delta} r_{t+j\cdot\Delta,\Delta}^2 \qquad (6.5)$$

In light of Eq. (6.5), Barndorff-Nielsen and Shephard (henceforth BNS, 2002, [8]) and ABDL [4] show that the *realized volatility* estimator is the sum of squared intraday returns on a given sampling frequency. For each day d and sampling frequency $1/m$, we compute:

$$RV^{d,m} = \sum_{i=1}^{m} r_{i,m}^2 \qquad (6.6)$$

As highlighted by Chevallier and Sévi (2011, [20]), CO_2 emissions futures contracts constitute new commodity assets against which energy traders need to develop hedging strategies. Thus, there lacks a fundamental body of knowledge as to what is the appropriate sampling frequency for this specific CO_2 intraday data. As in ABDL [3, 4], we use volatility signature plots[12] to estimate the range of sampling frequencies where the volatility is strongly increasing, indicating the increasing presence of microstructure noise. This phenomenon emerges from market microstructure problems, whose main examples are the existence of a bid-ask spread, non-synchronous trading, etc.

In view of Fig. 6.3, it appears that the choice of 15-minute returns should allow to minimize the impact of the microstructure noise, while ensuring for each day a sufficient number of observations. The use of 15-minute returns for the ECX carbon tick data also appears as a conservative choice compared to 5-minute sampling usually

---

[12]In volatility signature plots, the realized volatility measure described in Eq. (6.6) is computed and plotted at different sampling frequencies.

Fig. 6.3 Volatility signature plot for the ECX December 2008 futures contract from January 2, 2008 to December 15, 2008 with sampling frequencies ranging from 60 to 1,500 seconds

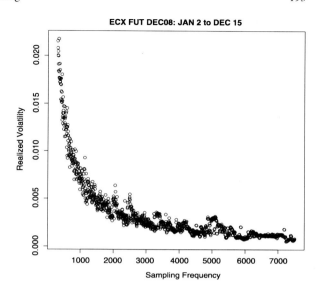

retained as the rule for other major equity and foreign currency exchange markets. This choice is motivated by practical reasons in order to minimize the impact of microstructure noise.[13]

Figure 6.4 plots the daily realized volatility of the ECX December 2008 and December 2009 futures contracts estimated with a 15-minute sampling frequency from January 2 to December 15 for each year. Summary statistics for the realized volatility measures may be found in Table 6.1. We observe that the daily realized volatility series present non-zero skewness and excess kurtosis. These descriptive statistics therefore reveal a 'fat tailed' leptokurtic distribution for the ECX CO_2 emissions futures contracts of maturity December 2008 and December 2009.[14]

6.6.2 Regression Analysis

Using intraday data, we examine the Samuelson hypothesis by regressing the natural logarithm of the realized volatility of the CO_2 futures on a constant and the number of days remaining to maturity:

$$rv_t = \alpha + \beta TTM_t + \varepsilon_t \qquad (6.7)$$

where rv_t is the natural logarithm of the ECX December 2008 (December 2009) futures realized volatility, α is a constant, and TTM_t represents the number of days to

---

[13] See Chevallier and Sévi (2010, [19]) for further statistical tests to determine the optimal sampling frequency with CO_2 intraday data.

[14] See Chevallier and Sévi (2011, [20]) for more details concerning the empirical characteristics of realized volatility in the EU ETS.

Fig. 6.4 Daily realized
volatility of the ECX
December 2008 (*top panel*)
and December 2009 (*bottom
panel*) futures contracts
estimated with a 15-minute
sampling frequency from
January 2 to December 15 for
each year

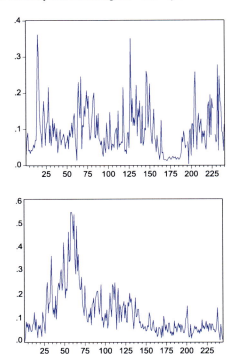

maturity. OLS regressions are computed using the Newey-West heteroskedasticity
consistent covariance matrix procedure.

6.6.3 Empirical Results

Estimation results are presented in Table 6.4. A *negative* and significant (at the 1%
level) β coefficient for the time-to-maturity may be observed for both futures contracts (regressions (5) and (6)). Contrary to the previous models, we find evidence in
support of the Samuelson hypothesis applied to the CO_2 allowance market by using
intraday data. The economic intuition behind this result is that the higher the frequency of the data observed, the higher the accuracy of volatility estimates for CO_2
futures prices. Compared to previous literature, the estimates of intraday volatility
based on realized measures are more accurate than the estimates based on daily data
which are used in Benz and Trueck (2009, [9]) and Oberndorfer (2009, [34]), among
others. Note also that, compared to the net carry cost approach (regressions (1) and
(2)), regressions (5) and (6) based on realized volatility modeling convey a higher
explanatory power, as judged by the adjusted R-squared.

Let us summarize the main results obtained so far. In order to examine the presence of the Samuelson hypothesis (that volatility increases closer to maturity) in
carbon futures markets, we rely on three different econometric methodologies and

rv_t	DEC08 (5)	DEC09 (6)
α	-0.0194^{***}	-0.2955^{***}
	(0.0056)	(0.0555)
β	-0.0137^{***}	-0.0022^{***}
	(0.0009)	(0.0006)
Diagnostic tests		
R-squared	0.0933	0.2102
F-Statistic	0.0000	0.0000
Durbin-Watson	1.8075	1.6923
Ljung-Box	0.8590	0.8339
White test	0.2082	0.1891
AIC	2.0253	0.4592
SC	2.0543	0.4878
N	240	244

Table 6.4 Testing the Samuelson hypothesis in the CO_2 allowance market using realized volatility

Note: Newey-West robust standard errors are presented in parentheses below coefficient estimates. ***, **, * indicate statistical significance at, respectively, the 1%, 5% and 10% levels. rv_t is the realized volatility estimator of the ECX December 2008 (*DEC*08) and December 2009 (*DEC*09) futures contracts. The model estimated is described in Eq. (6.7)

find that only based on the realized volatility approach is the hypothesis supported. Indeed, the net carry cost approach and GARCH modeling do not provide any (or rather weak) evidence in support of the presence of a Samuelson effect in the CO_2 futures prices. However, the use of high-frequency data shows that the daily (sum) volatility of the CO_2 futures prices increase near the expiration date, supporting the Samuelson hypothesis. Thus, the higher the frequency of the data observed, the higher the accuracy of volatility measures when dealing with carbon assets. The superiority of high frequency data in measuring the volatility of asset prices has been shown by other researchers (see ABDL [3, 4]).

Next, we proceed to sensitivity tests of these results.

6.6.4 Sensitivity Tests

6.6.4.1 Spot Price Volatility

First, we check the robustness of our estimates when the spot price volatility is included as an exogenous regressor (Bessembinder et al. 1996, [11]).

$$rv_t = \alpha + \beta TTM_t + \phi VolaSpot_t + \varepsilon_t \qquad (6.8)$$

where *VolaSpot$_t$* is the change in the spot price volatility (computed as one standard-deviation). This strategy allows to investigate the effect of the information flow, by comparing the results obtained when regressing the daily futures realized volatility on days to maturity with those obtained when the spot price volatility is also

Table 6.5 Testing the Samuelson hypothesis in the CO_2 allowance market using realized volatility and spot price volatility as an exogenous regressor

rv_t	DEC08	DEC09
	(7)	(8)
α	-0.0195^{***}	-0.4297^{***}
	(0.0049)	(0.0597)
β	-0.0078^{***}	-0.0038^{***}
	(0.0006)	(0.0006)
ϕ	0.0094	0.0079^{***}
	(0.0665)	(0.0027)
Diagnostic tests		
R-squared	0.0856	0.3185
F-Statistic	0.0000	0.0000
Durbin-Watson	1.6783	1.5845
Ljung-Box	0.8700	0.7510
White test	0.2925	0.2060
AIC	2.3955	0.5915
SC	2.4245	0.6202
N	240	244

Note: Newey-West robust standard errors are presented in parentheses below coefficient estimates. ***, **, * indicate statistical significance at, respectively, the 1%, 5% and 10% levels. rv_t is the realized volatility estimator of the ECX December 2008 (*DEC*08) and December 2009 (*DEC*09) futures contracts. The model estimated is described in Eq. (6.8)

included as a control variable. If the 'negative covariance' condition is the most significant factor behind the Samuelson hypothesis, then the coefficient on the days to maturity should remain negative and significant despite the inclusion of spot price volatility (Duong and Kalev 2008, [24]).

Estimation results are shown in Table 6.5. ϕ is not statistically significant in regression (7). However, for the December 2009 futures contract, ϕ is statistically significant at the 1% level and positive (regression (8)), which indicates that the volatility of spot prices has explanatory power over the 2009 futures price volatility in the CO_2 allowance market (note again the higher associated value of the adjusted R-squared).

In both regressions (7) and (8), it is noteworthy to remark that the sign of β remains negative and significant (at the 1% level) despite the inclusion of the *VolaSpot*$_t$ variable. These estimates show the robustness of the results identified in Eq. (6.7). With reference to the background discussion, the maturity effect in carbon futures using intraday data seems to be present even without the assumption on the information flow.

6.6.4.2 Seasonality and Liquidity

We control here for the potential effects of seasonality and liquidity on the time-to-maturity effects in carbon futures contracts. Indeed, the results derived previously may be driven by exogenous factors such as seasonal patterns or liquidity effects.

For instance, the Samuelson effect detected in intraday data may be due to a higher trading frequency as the carbon futures contracts approach their expiration date.[15]

To control for the effects of seasonality, we include dummy variables for the months of the year, which accommodate potential seasonalities in the information flow (Bessembinder et al. 1996, [11]). For example, the March dummy variable is composed of 1 during the period going from March 1 to March 31, and 0 thereafter.[16]

With regard to liquidity effects, we include the volumes exchanged for each carbon futures contract as an additional exogenous regressor on daily realized volatility. The model estimated thus becomes (Duong and Kalev 2008, [24]):

$$rv_t = \alpha + \beta TTM_t + \delta NT_t + \gamma_1 Feb_t + \gamma_2 Mar_t + \gamma_3 Apr_t + \gamma_4 May_t + \gamma_5 Jun_t \quad (6.9)$$

$$+ \gamma_6 Jul_t + \gamma_7 Aug_t + \gamma_8 Sep_t + \gamma_9 Oct_t + \gamma_{10} Nov_t + \gamma_{11} Dec_t + \varepsilon_t \quad (6.10)$$

with NT_t the number of trades, and Feb_t, Mar_t, Apr_t, May_t, Jun_t, Jul_t, Aug_t, Sep_t, Oct_t, Nov_t, Dec_t the monthly dummy variables.

Figure 6.5 shows the volumes exchanged in intraday data for the ECX futures contracts of maturity December 2008 and December 2009.

Out of a total volume exchanged of 1,842 and 3,947 million ton for, respectively, the December 2008 and December 2009 futures contracts,[17] we may observe visually that the highest volumes were exchanged on October 23, 2008 (36 million ton) and on February 19, 2009 (35 million ton). While the latter figure traditionally refers to the compliance period on the CO_2 allowance market (see Chevallier et al. 2009, [16] for more details), the former figure constitutes a record high when firms were selling allowances in exchange for cash in the midst of the 'credit-crunch' crisis. This volume analysis does not reveal, however, that the highest frequency of trading is recorded when maturity is approaching for carbon futures contracts.

Estimation results are provided in Table 6.6. The maturity effect detected in intraday data for carbon futures is robust to the introduction of exogenous regressors for seasonality and liquidity, as β remains significant (at the 1% level) and negative in regressions (9) and (10).

Concerning seasonality patterns, almost all monthly dummy variables are found to be significant (except February and April for the December 2008 contract, and August for the December 2009 contract).[18] The χ^2 statistics reported in the bottom

---

[15]Clark (1973, [22]) studies the relationship between volume and price volatility. The 'Mixture of Distribution Hypothesis' (MDH) states that trading volume and price volatility depend on the information flow on the market. As the rate of arrival of new information is not constant, price changes and volumes of trade tend to be stochastic in nature.

[16]Note the January dummy variable is omitted as standard procedure.

[17]Point Carbon analysts estimate that the volume of transactions on the European carbon market has been growing from 262 million ton in 2005 to 809 million ton in 2006, 1,455 million ton in 2007, 2,713 million ton in 2008, and 5,016 million ton in 2009. These estimates account for exchange-based trading as well as over-the-counter trading of emissions rights.

[18]Note we present the full model estimates here. Reduced-form estimates (i.e. only with significant variables) do not change qualitatively the results, as TTM_t remains significantly negative.

Fig. 6.5 Volumes exchanged (in 1,000 ton of CO_2) in intraday data for the ECX December 2008 (*top panel*) and December 2009 (*bottom panel*) futures contracts from January 2 to December 15 for each year

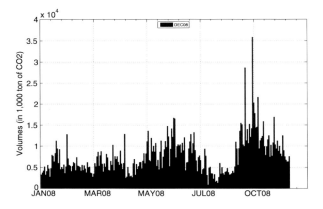

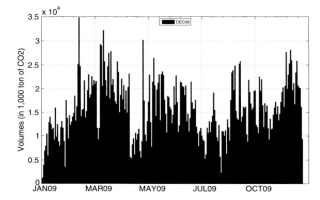

lines of Table 6.6 indicate that the hypothesis that the eleven estimated coefficients on the monthly indicators jointly equal to zero can be rejected for both futures contracts.

Therefore, we detect significant seasonalities in volatility for the carbon futures market. As might be expected, these seasonalities in volatility may be attributed to compliance cycles, as the γ coefficients for carbon futures tend to be highest during March-April. Note for the December 2008 futures contract, the γ coefficients tend to be highest later during that year due to the delayed impact of the 'credit crunch' (see comment above).

Concerning the liquidity effect, δ is not found to be significant in regression (9).[19] However, in regression (10), the variable NT_t impacts significantly (at the 1% level) and positively the realized volatility measure in logarithmic form for the December 2009 contract. This effect may certainly be understood by looking at Fig. 6.5 where, compared to the December 2008 contract, the volumes exchanged seem to

---

[19]We estimated regressions of volatility on the square root $\sqrt{NT}$ of the number of trades, without changing the result. We also investigated the sensitivity of test results to the use of the logarithmic transformation $\log(NT)$ of the number of trades, and found that conclusions are wholly unaffected.

Table 6.6 Testing the Samuelson hypothesis in the CO_2 allowance market using realized volatility with seasonal dummy variables and number of trades as exogenous regressors

rv_t	DEC08 (9)	DEC09 (10)
α	−0.0919	−0.8835
	(0.0717)	(0.6420)
β	−0.0100***	−0.0016***
	(0.0009)	(0.0004)
δ	0.0001	0.0001***
	(0.0001)	(0.0001)
γ_1	0.0514	0.2429**
	(0.2139)	(0.1023)
γ_2	−0.4104**	0.4943***
	(0.1899)	(0.0651)
γ_3	−0.1715	0.4948***
	(0.2129)	(0.0989)
γ_4	−0.6256***	0.2646***
	(0.2381)	(0.0742)
γ_5	−0.0560***	0.1058*
	(0.0160)	(0.0644)
γ_6	−0.9736***	0.1490***
	(0.2287)	(0.0483)
γ_7	−0.9355***	−0.0795
	(0.1660)	(0.0646)
γ_8	−0.0709***	−0.2721***
	(0.0291)	(0.0733)
γ_9	−0.7369***	−0.3331***
	(0.1261)	(0.0811)
γ_{10}	−0.9141***	−0.4101***
	(0.2630)	(0.0822)
γ_{11}	−0.8350***	−0.4528***
	(0.1206)	(0.0787)

Note: Newey-West robust standard errors are presented in parentheses below coefficient estimates. ***, **, *indicate statistical significance at, respectively, the 1%, 5% and 10% levels. rv_t is the realized volatility estimator of the ECX December 2008 (*DEC*08) and December 2009 (*DEC*09) futures contracts. The model estimated is described in Eq. (6.9)

Diagnostic tests		
R-squared	0.0889	0.5960
F-Statistic	0.0000	0.0000
Durbin-Watson	1.5003	1.3355
Ljung-Box	0.7400	0.7020
White test	0.4035	0.2748
AIC	1.6704	0.1715
SC	1.8589	0.0148
N	240	244
Indicators $= 0$	990.5600	216.0701
χ^2 (*p*-value)	0.0000	0.0000

be significantly higher. Note that the effect of TTM_t still remains persistent in that case.

Thus, our empirical evidence regarding the Samuelson hypothesis appears robust even after controlling for the effects of seasonality and liquidity. As in Bessembinder et al. (1996, [11]), similar regressions of daily volatility on days to expiration, spot volatility and monthly indicators yield the same results. Overall, we may conclude that high-frequency data allow to detect statistically the presence of time-to-maturity effects in the ECX December 2008 December 2009 futures contracts.

6.7 Summary

This chapter investigates the 'Samuelson hypothesis' in the CO_2 allowance market by using daily and intraday data. We define the Samuelson effect as an increase in the standard deviation of the futures contract price changes as the time-to-maturity approaches. Testing this relationship has indeed important implications in terms of margin setting, hedging, and option pricing in futures markets.

The main findings may be summarized as follows. First, following Bessembinder et al. (1995, [10]), changes in the net carry cost for CO_2 prices are not statistically significantly and negatively affected by changes in spot prices. Thus, the non validity of the 'negative covariance' hypothesis apparently yields to reject the presence of maturity effects in the CO_2 futures prices. Second, we further investigate whether the Samuelson hypothesis holds for CO_2 allowances by fitting the AR(1)-GARCH(1, 1) model to the daily returns of the ECX December 2008 and December 2009 futures contracts, and by adding the time-to-maturity as an exogenous regressor in the variance equation. Similarly to the net carry cost approach, these results are inconsistent with the presence of maturity effects in the CO_2 futures prices.

Third, we use intraday data to compute the realized volatility of the ECX December 2008 and December 2009 futures contracts. The regressions based on the natural logarithm of the realized volatility measure show a *negative* and statistically significant effect (at the 1% level) of the number of days to maturity, which demonstrates the existence of expiration effects in the CO_2 futures price series. The latter result is checked for robustness by including the volatility of spot prices in the regression. For the December 2008 contract, the addition of the spot volatility proxy has little effect on previous estimates associated with time to expiration. For the December 2009 contract, besides the stability of the coefficients for the time-to-maturity variable, we find a positive and statistically significant effect (at the 1% level) concerning the ability of spot prices to explain futures prices. The last sensitivity test deals with the potential effects of seasonality and liquidity. Although each carbon futures contract is characterized by seasonal patterns in volatility (captured with the inclusion of monthly dummy variables), inference regarding whether the Samuelson hypothesis holds is robust to these seasonalities. Finally, these empirical results are robust to the inclusion of the number of trades that proxy for the liquidity on the carbon futures market.

Using intraday data, we may conclude that we support the presence of the Samuelson maturity effect in the CO_2 futures prices. This property of (daily) GARCH modeling *vs.* (intraday) realized volatility modeling is widely documented in previous literature on equity markets [3, 4, 30]. Indeed, GARCH models track much better the broad temporal movements in the volatilities for lower frequency variations, and their accuracy tends to perform poorly at higher frequencies. This is due to the superiority of realized measures in estimation.

Appendix: Statistical Techniques to Detect Instability in the Volatility of Carbon Prices

This section develops additional statistical tests to detect structural breaks in the volatility of EUAs (see [17]), based on the econometric methodology detailed in [40] and [41]. To do so, we compute a measure of volatility for European Union Allowances, based on the EGARCH model. We then proceed with developing various kinds of structural break tests in order to analyze the volatility dynamics overtime.

EGARCH Modeling We study the conditional standard deviation extracted from an EGARCH model [33]:

$$\log(\sigma_t^2) = \omega + \sum_{i=1}^{p} \alpha_i \left| \frac{\varepsilon_{t-i}}{\sigma_{t-i}} \right| + \sum_{j=1}^{q} \beta_j \log(\sigma_{t-j}^2) + \sum_{k=1}^{r} \gamma_k \frac{\varepsilon_{t-k}}{\sigma_{t-k}} \tag{6.11}$$

with σ_t^2 the conditional variance, which is function of a constant term ω, the ARCH term ε_{t-i}^2, and the GARCH term σ_{t-j}^2. γ tests for the presence of the leverage effect. The corresponding volatility measure for EUAs is pictured in Fig. 6.6 during the period going from March 9, 2007 to March 31, 2009 (i.e. a sample of 529 daily observations).

The data for this section is available at: http://sites.google.com/site/jpchevallier/publications/books/springer/data_chapter6_appendix.csv.

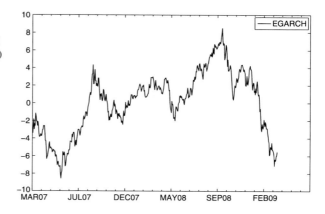

Fig. 6.6 Conditional standard deviation for EUAs extracted from the EGARCH model estimated in Eq. (6.11)

Table 6.7 R Code: OLS/Recursive based CUSUM Processes

```
1  path<-"C:/"
2  setwd(path)
3  library(strucchange)
4  data=read.csv("data_chapter_appendix.csv",sep=",")
5  attach(data)
6  garch=data[,2]
7  garch.model <- garch~1
8  garch.cusum <- efp(garch.model, type = "OLS-CUSUM")
9  plot(garch.cusum)
10 garch.recusum <- efp(garch.model, type = "Rec-CUSUM")
11 plot(garch.recusum)
```

OLS/Recursive Based CUSUM Processes   First, to check for structural changes in the model, we use the OLS-based CUSUM process [36] which contains cumulative sums of standardized residuals.:

$$W_n^0(t) = \frac{1}{\hat{\sigma}\sqrt{n}} \sum_{i=1}^{\lfloor nt \rfloor} \hat{u}_i \quad (0 \leq t \leq 1) \tag{6.12}$$

with $W_n^0(t)$ the limiting process for the standard Brownian bridge $W^0(t) = w(t) - tW(1)$, and $W(.)$ the standard Brownian motion. Under a single-shift alternative, the process should have a peak around the breakpoint. These processes are also computed by specifying recursive instead of OLS estimates. The R code is shown in Table 6.7.

Lines 1 to 6 load the relevant data and libraries. Lines 7 to 9 estimate the OLS-based CUSUM processes. Lines 10 and 11 compute and plot the recursive-based CUSUM processes.

Figure 6.7 shows the various processes together with their boundaries at an (asymptotic) 5% significance level. The EGARCH process clearly exhibit some instability with two distinct patterns: (i) a negative peak during July-December 2007 (which coincides with the end of Phase I and the design inefficiencies related to banking restrictions, see [2]), and (ii) a positive peak during February-March 2009 (which may be linked to the 2008 compliance event and the perceived increased scarcity of CO_2 allowances due to NAP II negotiations). These peaks exceed the boundaries and hence indicate structural breaks during these periods. Note that, in this illustrative example, the recursive-based CUSUM processes indicate a period of instability during most of the time spanned by our data sample (from July 2007 onwards).

Significance Testing   Second, we carry out significance tests under the null hypothesis of 'no structural break':

$$S_r = \max_t \frac{efp(t)}{f(t)} \tag{6.13}$$

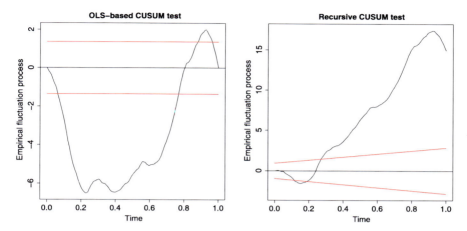

Fig. 6.7 OLS- (*left panel*) and Recursive- (*right panel*) based CUSUM processes for the EGARCH model

with S_r the test statistic for the residual-based processes estimated in the previous paragraph, $efp(t)$ the given empirical fluctuation process at time t, and $f(t)$ depends on the shape of the boundary (i.e. $b(t) = \lambda f(t)$). F-statistics are calculated for all potential change points (F_i for $k < \underline{i} \le i \le \bar{i} < n - k$), and the null hypothesis is rejected if any of those statistics gets too large.

We retain two possibilities to aggregate the series of F-statistics into a test statistic [6, 7]:

$$\sup F = \sup_{\underline{i} \le i \le \bar{i}} F_i \tag{6.14}$$

$$\text{ave } F = \frac{1}{\bar{i} - \underline{i} + 1} \sum_{i=\underline{i}}^{\bar{i}} F_i \tag{6.15}$$

with $\sup F$ and $\text{ave } F$ the supremum and average of the F-statistics, respectively. Under the null hypothesis of no structural break, boundaries can be computed such that the asymptotic probability that the supremum (or the mean) of the statistics F_i (for $\underline{i} \le i \le \bar{i}$) exceeds this boundary is α. The R code is given in Table 6.8.

As previously, Lines 1 to 7 load the model. Lines 8 and 9 compute the F-statistic for all potential structural breaks, along with the p-value. The output can be seen in Fig. 6.8.

The dashed line in each panel of Fig. 6.8 represents the observed mean of the F-statistics, where the boundaries have been set up for the average instead of the supremum. As the F-statistics of the EGARCH process regularly cross its boundaries, there is statistical evidence for structural breaks (at the 5% level). The process has clear peaks in April 2006, October 2007, and October 2008, which mirrors the results obtained from the analysis by OLS- and Recursive-based CUSUM processes. Besides, we note the large p-value obtained near the end of the sample

Table 6.8 R Code: F-Statistic for structural breaks

```
1 path<-"C:/"
2 setwd(path)
3 library(strucchange)
4 data=read.csv("data_chapter_appendix.csv",sep=",")
5 attach(data)
6 garch=data[,2]
7 garch.model <- garch~1
8 fs.garch.sup <- Fstats(garch.model)
9 plot(fs.garch.sup)
```

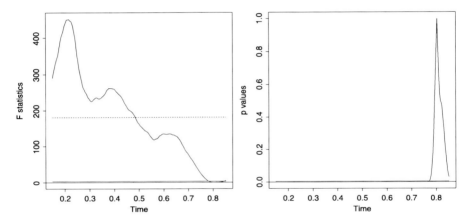

Fig. 6.8 F-Statistic (*left panel*) and p-value (*right panel*) for structural breaks in the EGARCH model

during February-March 2009 (which may be linked to the 2008 compliance event as commented above).

Monitoring Structural Breaks   Third, we aim at monitoring the evolution of our volatility series overtime. Monitoring has been developed for recursive [21] and moving [28] estimates tests. When new observations arrive, estimates are computed sequentially from all available data (historical sample plus newly arrived data), and compared to the estimate based only on the historical sample. The hypothesis of no structural break is rejected if the difference between these two estimates gets too large:

$$y - i = x_i^\top \beta_i + u_i \quad (i = 1, \dots, n, n + 1, \dots) \tag{6.16}$$

i.e., we expect new observations to arrive after time n (when the monitoring begins). The sample $\{(x1, y1), \dots, (x_n, y_n)\}$ is called the historic sample, and the corresponding time period $1, \dots, n$ the history period. In what follows, we focus on

Table 6.9 R Code: Monitoring structural breaks

```
1  path<-"C:/"
2  setwd(path)
3  library(strucchange)
4  data=read.csv("data_chapter_appendix.csv",sep=",")
5  attach(data)
6  garch=data[,2]
7  garch.model <- garch~1
8  serie<-ts(data, start=c(1,1),frequency=7)
9  serie<-ts(data, start=c(1,1),end=c(101,1),frequency=7)
10 garch.mefp.2008 <- mefp(garch.model, type = "ME",
   data = serie, alpha = 0.05)
11 garch.mefp.2008.fe <- mefp(garch.model,
   type = "fluctuation", data = serie, alpha = 0.05)
12 serie<-ts(data, start=c(1,1),end=c(135,1),frequency=7)
13 garch.mefp.2008 <- monitor(garch.mefp.2008)
14 garch.mefp.2008.fe <- monitor(garch.mefp.2008.fe)
15 plot(garch.mefp.2008)
16 plot(garch.mefp.2008.fe)
```

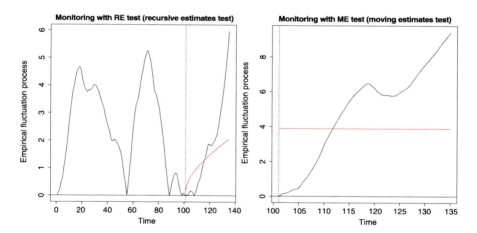

Fig. 6.9 Monitoring EUA data with recursive (*left panel*) and moving (*right panel*) estimates tests for the EGARCH model

the two-year sample gathered for the volatility time-series of EUAs (March 2007 to March 2009). The R code is provided in Table 6.9.

Lines 1 to 9 load the model and set the frequency of the time-series. Lines 10 to 12 specify the length of the historical sample. Lines 13 to 16 launch the monitoring, and plot the empirical fluctuation process. Results for both recursive and moving estimates are shown in Fig. 6.9.

According to recursive estimates, the breakpoint is detected during September-October 2008, which corresponds to a period of high liquidity on the European carbon market in the midst of the financial crisis. Indeed, market operators were selling allowances in exchange of cash during that period, which may contribute to explain why we identify a breakpoint in the volatility of the EUA time series [17]. With moving estimates, we obtain a breakpoint at the beginning of the sample, i.e. during the 2006 compliance event [18]. Thus, we are able to isolate precisely the events that impact the volatility of carbon prices based on this econometric methodology, where we have not one history estimate that is being compared with the new moving estimates, but we have a history process.

References

1. Adrangi B, Chatrath A (2003) Non-linear dynamics in futures prices: evidence from the coffee, sugar and cocoa exchange. Appl Financ Econ 13:245–256
2. Alberola E, Chevallier J (2009) European carbon prices and banking restrictions: evidence from Phase I (2005–2007). Energy J 30:51–80
3. Andersen TG, Bollerslev T, Diebold FX, Ebens H (2001) The distribution of stock return volatility. J Financ Econ 61:43–76
4. Andersen TG, Bollerslev T, Diebold FX, Labys P (2003) Modeling and forecasting realized volatility. Econometrica 71:579–625
5. Anderson RW, Danthine J (1983) The time pattern of hedging and the volatility of futures prices. Rev Econ Stud 50:249–266
6. Andrews DWK (1993) Tests for parameter instability and structural change with unknown change point. Econometrica 61:821–856
7. Andrews DWK, Ploberger W (1994) Optimal tests when a nuisance parameter is present only under the alternative. Econometrica 62:1383–1414
8. Barndorff-Nielsen O, Shephard N (2002) Econometric analysis of realized volatility and its use in estimating stochastic volatility models. J R Stat Soc Ser B 64:253–280
9. Benz E, Trück S (2009) Modeling the price dynamics of CO_2 emission allowances. Energy Econ 31:4–15
10. Bessembinder H, Coughenour JF, Seguin PJ, Smoller MM (1995) Mean reversion in equilibrium asset prices: evidence from the futures term structure. J Finance 50:361–375
11. Bessembinder H, Coughenour JF, Seguin PJ, Smoller MM (1996) Is there a term structure of future volatilities? Reevaluating the Samuelson hypothesis. J Deriv 4:45–58
12. Berndt E, Hall B, Hall R, Hausman J (1974) Estimation and inference in nonlinear structural models. Ann Econ Soc Meas 3:653–665
13. Board J, Sutcliffe C (1990) Information volatility, volume, and maturity: an investigation of stock index futures. Rev Futures Mark 9:532–549
14. Brennan M (1958) The supply of storage. Am Econ Rev 48:50–72
15. Chen YJ, Duan JC, Hung MW (1999) Volatility and maturity effects in the Nikkei Index Futures. J Futures Mark 19:895–909
16. Chevallier J (2009) Carbon futures and macroeconomic risk factors: a view from the EU ETS. Energy Econ 31:614–625
17. Chevallier J (2011) Detecting instability in the volatility of carbon prices. Energy Econ 33:99–110
18. Chevallier J, Ielpo F, Mercier L (2009) Risk aversion and institutional information disclosure on the European carbon market: a case-study of the 2006 compliance event. Energy Policy 37:15–28
19. Chevallier J, Sévi B (2010) Jump-robust estimation of realized volatility in the EU emissions trading scheme. J Energy Markets 3:49–67
20. Chevallier J, Sévi B (2011) On the realized volatility of the ECX CO_2 emissions 2008 futures contract: distribution, dynamics, and forecasting. Ann Finance 7:1–29

21. Chu CSJ, Stinchcombe M, White H (1996) Monitoring structural change. Econometrica 64:1045–1065
22. Clark PK (1973) A subordinated stochastic process model with finite variance for speculative prices. Econometrica 41:135–155
23. Daskalakis G, Psychoyios D, Markellos RN (2009) Modelling CO_2 emission allowance prices and derivatives: evidence from the European trading scheme. J Bank Finance 33:1230–1241
24. Duong HN, Kalev PS (2008) The Samuelson hypothesis in futures markets: an analysis using intraday data. J Bank Finance 32:489–500
25. Haff I, Lindqvist O, Leiland A (2008) Risk premium in the UK natural gas forward market. Energy Econ 30:2420–2440
26. Kalev PS, Duong HN (2008) A test of the Samuelson hypothesis using realized range. J Futures Mark 28:680–696
27. Kolb R (1991) Understanding futures markets. Kolb Publishing Company, Miami
28. Leisch F, Hornik K, Kuan CM (2000) Monitoring structural changes with the generalized fluctuation test. Econom Theory 16:835–854
29. Martens M, van Dijk D (2007) Measuring volatility with the realized range. J Econom 138:181–207
30. McMillan DG, Speight AEH (2004) Intra-day periodicity, temporal aggregation and time-to-maturity in FTSE-100 index futures volatility. Appl Financ Econ 14:253–263
31. Movassagh N, Modjtahedi B (2005) Bias and backwardation in natural gas futures prices. J Futures Mark 25:281–308
32. Mu X (2007) Weather, storage, and natural gas price dynamics: fundamentals and volatility. Energy Econ 29:46–63
33. Nelson DB (1991) Conditional heteroskedasticity in asset returns: a new approach. Econometrica 59:347–370
34. Oberndorfer U (2009) EU emission allowances and the stock market: evidence from the electricity industry. Ecol Econ 68:1116–1126
35. Paolella MS, Taschini L (2008) An econometric analysis of emission allowance prices. J Bank Finance 32:2022–2032
36. Ploberger W, Kramer W (1992) The CUSUM test with OLS residuals. Econometrica 60:271–285
37. Samuelson PA (1965) Proof that properly anticipated prices fluctuate randomly. Ind Manage Rev 6:41–49
38. Thomakos DD, Wang T (2003) Realized volatility in the futures markets. J Empir Finance 10:321–353
39. Working H (1949) The theory of the price of storage. Am Econ Rev 39:1254–1262
40. Zeileis A, Kleiber C, Krämer W, Hornik K (2003) Testing and dating of structural changes in practice. Comput Stat Data Anal 44:109–123
41. Zeileis A (2006) Implementing a class of structural change tests: an econometric computing approach. Comput Stat Data Anal 50:2987–3008

Solutions

A.1 Problems of Chap. 2

Problem 2.1 (Unit Root Tests for Carbon Prices)

(a) In Fig. 2.16, the logreturns seem stationary. Concerning the raw time series, we may observe a stationary component during the second half of the sample. During the first half of the sample, on the contrary, we can see that the time series is not stationary. This non stationarity does not seem to follow a stochastic process. Instead, the downside in the price series may be explained by a downward trend. This trend may be explained by specific events occurring on the carbon market (such as compliance events or banking restriction effects).

(b) In Fig. 2.17, the ACF of the raw time series does not provide us with any useful information. Recall that ACF are especially useful to investigate following the estimation of a given model, in order to check the residual properties of the time series. From the graph of the logreturns, we learn that the EUA ECX Futures contract may be configured with an AR(1) process.

(c) Tables 2.15 and 2.16 constitute a classic example of nonstationary raw time series and stationary logreturns. In Table 2.15, the ADF and PP tests cannot reject the null hypothesis of unit root for the EUA raw time series, while the KPSS test rejects the null hypothesis of stationarity. Therefore, the 3 tests point to the same conclusion that the time series of raw carbon prices is not stationary. In Table 2.16, on the contrary, the ADF and PP tests strongly rejects the null hypothesis of unit root for the logreturns, while the KPSS test accepts the null hypothesis of stationarity. Therefore, the time series of carbon prices is integrated of order 1 ($I(1)$). There do not seem to be any bias in the methodology used, as the EUA time series does not respond to any trend or time dependence (as highlighted in the reply to question (a)). Maybe there is a need to conduct a stationarity test with the presence of one structural break during the first half of the sample, instead of the classic unit root tests.

J. Chevallier, *Econometric Analysis of Carbon Markets*,
DOI 10.1007/978-94-007-2412-9, © Springer Science+Business Media B.V. 2012

Problem 2.2 (Calibration of Mean-Reversion Models for Carbon Prices)

(a) The estimation of an AR(1) model will do the trick.
(b) In Table 2.17, we notice that the configuration of an AR(1) process for the lo-
    greturns of the EUA time series appears satisfactory. Indeed, the coefficient for
    the variable $DIFFEUA(-1)_t$ is statistically significant at the 10% level.
(c) In Table 2.18, we conclude that the residuals are not autocorrelated according
    to the Box-Pierce test (as the p-value of the test is higher than 5%). However,
    we notice the presence of strong ARCH effects. The residuals are not normally
    distributed, which is a standard characteristic of financial time series.
(d) The ARCH effects could be taken into account by specifying a GARCH model.
    Concerning the normality assumption, we could replace the normal distribution
    by a density presenting fat tails (e.g. the Normal Inverse Gaussian or General-
    ized Error Distribution).

A.2 Problems of Chap. 4

Problem 4.1 (The EUA-CER Spread)

(a) In Fig. 4.17, the left panel contains the raw time series of EUAs, CERs, as
    well as the EUA-CER spread. The right panel contains the time series of EUAs
    and CERs transformed in logarithm. This logarithmic transformation allows
    to remove size effects between time series (for instance when they are not
    expressed with the same currency or quantity), without changing their char-
    acteristics in terms of stationarity. The logarithmic transformation can be pre-
    ferred to raw time series for many econometric applications. In the right panel,
    the EUA-CER spread cannot be expressed in logs, as it contains negative
    values.
(b) In Table 4.21, comment on the non-normal characteristics (in terms of skew-
    ness and kurtosis, see Chap. 1) of EUAs and CERs for the raw and logarithmic
    transformation of the time series.
(c) Tables 4.22 and 4.23 deal with the Johansen cointegration tests between EUAs
    and CERs. The maximum eigenvalue and the trace statistic tests cannot identify
    any cointegration relationship between the two time series, at the 5% signifi-
    cance level. Hence, we accept the null hypothesis that the cointegration space
    is equal to $r = 0$, as no value for the test statistics is found to be superior
    to the critical values of the test. This result is not surprising, since we have
    discussed in Sect. 4.2.2.4 the necessity to take into account structural breaks
    during the period to uncover a cointegration relationship between EUAs and
    CERs.
(d) In Tables 4.24 to 4.26, if you have correctly answered to Question (c), then
    there is no need to comment upon the ECM since the model is mis-specified
    (i.e. there is no cointegration relationship between the two variables according

to the Johansen cointegration tests). This question is a trap designed to test your ability to detect potentially false econometric results.

(e) In Fig. 4.18, EUAs and CERs have been transformed to log-returns. The main difference with Fig. 4.17 is that the variables are stationary.

(f) In Table 4.27, comment similarly the properties of the transformed time series with respect to the normal distribution.

(g) In Tables 4.28 and 4.29, the VAR(4) model has been estimated based on the AIC and the results of the Portmanteau test (as the residuals were autocorrelated for the VAR(1) model). The results show that the CERs lagged 1 and 3 periods have a statistically significant impact on the EUA price. The EUA variable is affected by its own lags 2 and 4 at statistically significant levels. Concerning the CER variable, there is evidence of significant autoregressive parameters at lags 1 and 3. The CER price is negatively affected by the EUA variable at lag 3. Therefore, there is evidence of interrelationships between the two variables across the estimates of the VAR. We can consider that the VAR model is correctly specified (since we do not constrain all coefficients to be significant in a VAR).

(h) In Table 4.30, we can indeed conclude that the VAR(4) model has been well specified, as judged by the high p-value for the Portmanteau test.

(i) In Table 4.31, there is evidence of reverse causality in the Granger sense between the two variables.

(j) In Fig. 4.19, it is difficult to say which Cholesky ordering has been specified, since there is evidence of reverse causality (in the Granger sense). We can presume that the EUA variable is the least exogenous variable in this system, since it corresponds to the most important carbon price. The IRF functions have the standard 'hump' shape, with an initial decline in the time series following the shock (-0.15 standard deviation at lag 4), and then the effect of the shocks gradually disappear (at the horizon of lag 10) which is characteristic of the stationarity of the VAR.

(k) In Table 4.32, the variance equation shows significant GARCH effects for the EUA-CER spread. The constant term is not significant in the mean equation, but it is significant in the variance equation. The ARCH coefficient is not statistically significant. The sum of the ARCH and GARCH term is well below 1. Besides, we can conclude from the diagnostic tests that the residuals are not autocorrelated, and that there are no remaining GARCH effects.

(l) In Fig. 4.20, we can still notice one significant lag in the autocorrelation structure of the GARCH estimates according to the ACF.

(m) In Fig. 4.21, we can visually confirm that the innovations (top panel) and standardized innovations (bottom panel) of the GARCH model satisfy the required properties. More interestingly, we notice in the middle panel that there are peaks in the conditional standard deviation of the CER-EUA spread (as modeled by the GARCH process) in July 2008 and in December 2008, which certainly deserve more attention from an economic viewpoint.

A.3 Problems of Chap. 5

Problem 5.1 (Calibrating GARCH Models for Carbon Prices)

(a) In Table 5.9, we observe:

- Little serial correlation in the raw returns (p-value of the Ljung-Box test > 5% level).
- Significant serial correlation in the squared returns (p-value of the Ljung-Box test < 5% level).
- Evidence of heteroskedasticity (ARCH test < 5% level).

  These data features a compatible with the GARCH(p, q) estimation.

(b) In Table 5.10, the AR(1)-GARCH(1, 1) model is estimated. The AR(1) process is not significant in the mean equation, which points to a mis-specification problem in the configuration of the data generating process. The ARCH and GARCH terms are positive and significant at the 1% level in the variance equation, which means that the GARCH model is well estimated. The GARCH effect seems predominant in the CO_2 price series.

  In the diagnostic tests of Table 5.10, we observe:

- No correlation;
- The ARCH effects have been properly taken into account by the model.

  In Fig. 5.10, we observe:

- Little volatility clustering;
- Some patterns in the conditional standard deviation, which may be related to market events. Therefore, we could add dummy variables for structural breaks, as explained in Chap. 2.

(c) In Tables 5.11 and 5.12, we observe an improvement over the initial specification, as some (but not all) of the leverage parameters are statistically significant in the variance equation. The same comments as for the GARCH model apply concerning the diagnostic tests of the EGARCH and TGARCH models. Concerning the choice of the distribution, the Student t seems to provide the best goodness-of-fit. Finally, we can still notice that the AR(1) is not significant in the mean equation of both models, indicating a potential mis-specification problem.

Index

Symbols
F-statistic, 189, 203
p-value, 203

A
AB32, 2
Abatement cost, 169
ACF, 25, 35, 39, 43, 86, 109, 128, 144, 211
Acid Rain Program, 6
ADF, 21, 48, 164
ADF test, 22, 72
Adjusted R-Squared, 35
AIC, 83, 140, 164, 189, 211
Allocation, 11, 99
Allocation rule, 5, 148
Allowance supply, 66
Ang and Bekaert (2002), 137
AR, 82
AR(1), 23, 49, 190, 191
AR(1)-GARCH(1, 1), 190, 200
AR(1)-GARCH(1,1), 212
ARA, 31, 128, 161, 163, 167
Arbitrage, 132, 184
ARCH, 56, 72, 122, 164, 190, 191, 211
ARIMA, 22
ARMA, 56, 66, 82, 164, 176
Asset management, 171
Auctioning, 5
Australia, 3
Autocorrelation, 26, 91, 144, 190, 211
AutoRegressive Integrated Moving Average, 22

B
Bai and Perron, 27
Banking, 6, 9

Banking restrictions, 8, 182
Bayesian Information Criterion, 28
BDS test, 76
BEKK, 45, 46
BEKK MGARCH, 48
BERR, 161
Bessembinder and Lemmon's model, 156
BHHH, 55, 56, 142, 190
Bid-ask spread, 192
BIS, 161
Black-Scholes, 173
BlueNext, 13
Bollerslev-Wooldridge, 190
Bond market, 66
Bonds, 60, 61
Borrowing, 6
Box-Jenkins, 164, 176, 190
Box-Pierce test, 26, 49
Breakpoint, 205
Brent price, 31, 127
Breusch-Godfrey, 109
Brokers, 12, 149
Business cycle, 60, 62

C
Calendar, 12
CAPM, 8
Carbon price risk, 169
Carbon-macroeconomy relationship, 71, 92, 100
CCC MGARCH, 48
CCGT, 169
CDC Climate Research, 42
CDM, 11, 105, 147
CDM EB, 105, 148
CDM Executive Board, 105, 148

J. Chevallier, *Econometric Analysis of Carbon Markets*,
DOI 10.1007/978-94-007-2412-9, © Springer Science+Business Media B.V. 2012

CER, 105, 107, 131, 138, 148, 149, 170, 210
CER-EUA spread, 131, 133
CGARCH, 65, 190
CH_4, 150
Chaos, 76
China, 3
Cholesky decomposition, 111, 115, 116, 142,
       211
CIF ARA, 170
CITL, 4, 11, 149
Clean Dark Spread, 37
Clean Spark Spread, 38
Climate policies, 1
CO_2 emissions, 60, 71
Coal, 30, 45
Coal price, 31, 127, 167
Cointegration, 107, 119, 139, 210
Combined Cycle Gas Turbines, 169
Commodities, 60, 63, 66, 100, 149, 151, 154,
       183, 189
Commodity markets, 182
Commodity pricing, 8
Compliance, 198
Compliance event, 14, 149, 206
Conditional standard deviation, 211
Conditional volatility, 181
Convenience yield, 8, 152, 187
COP/MOP, 1, 151
Cost of carry, 153, 186
Covariance matrix, 67
CPRS, 3
CRB, 63
Cross-correlations, 31, 39
CUSUM, 73, 202

D
DCC MGARCH, 48, 123
Default risk, 61
Delivery, 184
Descriptive statistics, 140
Diagnostic check, 144
Dickey-Fuller, 20
Directive 2003/87/CE, 10
Dividend yield, 60, 61, 66
Double accounting period, 12
Dow Jones Eurostoxx 50, 61
Dummy variable, 19, 20, 43, 128, 197, 200
Duration, 99, 134
Durbin-Watson, 189
Dynamic linear model, 23

E
EC, 105, 149
ECB, 161, 170

ECM, 210
ECX, 13, 14, 71, 170, 200
EEX, 13, 170
EGARCH, 65, 177, 190
Eigenvalue, 67
EITE, 3
Electricity, 30
Electricity price, 36, 153
Energy market, 66
Energy-Climate Package, 4
Engle ARCH test, 45, 49, 164, 191
Error-correction model, 117, 118, 120, 139
EU ETS, 147
EUA, 11, 14, 45, 71, 107, 131, 138, 149, 162,
       163, 170, 206, 210
EUA-CER spread, 138, 210
Euronext 100, 61
European Central Bank, 161
European Commission, 14, 66, 105, 149
Excess return, 60, 62, 64
Exhaustible resource, 6, 9
Exhaustion condition, 6
Expiration, 187, 195, 200
Expiration date, 181

F
F-statistic, 35
Factor model, 66
Factor-Augmented VAR, 68
Factors, 66
FAVAR, 68
Flexibility mechanism, 7
Forecasting, 181
Free lunch, 132
FTSE, 182
Fuel mix, 169
Fuel-switching, 11, 66, 159, 169
Futures, 181, 190, 196
Futures market, 200
Futures premium, 154, 155
Futures pricing, 183
Futures-spot bias, 151
Futures-spot structural model, 154

G
GARCH, 31, 56, 64, 66, 144, 164, 174, 181,
       182, 185, 190, 191, 195, 201, 202,
       211
GARCH-in-Mean, 65
Gas, 45
Gas price, 166
GED, 177
Generalized Error Distribution, 177
GHG, 1
Grandfathering, 4, 5, 148

Granger causality, 33, 115, 141, 211
GSCI, 63

H

Hedge ratio, 183
Hedger, 183
Hedging, 181, 200
Heteroskedasticity, 45, 91, 189, 194
HFC, 150
HFC-23, 150, 151
High-frequency data, 191, 195
Hotelling condition, 6
Hotelling rule, 8
Hotelling-CAPM model, 9

I

ICE, 31, 170
IET, 10
Implied volatility, 173
Import limit, 150
Impulse response function, 69, 116, 141
Industrial production, 71, 94, 100
Information criteria, 108
Information flow, 183, 195, 196
Innovations, 211
Instability, 73
Installations, 12
Institutional decisions, 19
Interest rate, 62, 66
International Emissions Trading, 10
International Transactions Log, 148
Intertemporal exchange ratio, 7
Intraday data, 181, 182, 186, 192–194, 196, 200
Inventory, 187
IRF, 69, 116, 141
ITL, 4, 11, 148, 149
ITL-CITL connection, 149
ITR, 7

J

Japan, 3
Jarque-Bera, 15, 72, 110
Jarque-Bera test, 49
JI, 11
Johansen cointegration test, 118, 139, 210
Junk bond yield, 60, 61, 64

K

Keenan test, 75
KPSS, 22, 48, 164
KPSS test, 22, 72
Kronecker, 46
Kurtosis, 15, 16, 110, 138, 193, 210
Kyoto Protocol, 1

L

Lag, 164
Lagrange multiplier, 45
Leverage effect, 64, 65, 201
Likelihood Ratio test statistic, 79
Linking Directive, 107, 150
Liquidity, 182, 184, 196, 197, 200, 205
Ljung-Box, 57, 72, 164
Ljung-Box test, 26, 189, 191
Ljung-Box-Pierce test, 26
Log-returns, 211
Logarithmic transformation, 210
LR test, 79, 81
LULUCF, 150

M

MAC, 5
Macroeconomic indicator, 62
Macroeconomic risk factor, 60, 64, 66
Macroeconomic shock, 60
Macroeconomic variable, 66
Macroeconomy, 60
Margin call, 183
Margin setting, 181, 183, 200
Marginal abatement cost, 5
Market fundamentals, 29
Market microstructure, 192
Markov Chain Monte Carlo Methods, 69
Markov-switching, 71, 88, 92, 97, 100, 138
Markov-switching VAR, 134
Matrix of cross-correlations, 33, 128
Maturity, 152, 183, 200
Maximum eigenvalue test, 118, 210
Maximum likelihood, 93, 121, 134
MCMC, 69
MDH, 197
Mean reversion, 49, 64, 187
Mean-variance optimization, 170
MGARCH, 45
MGGRA, 2
Microstructure, 133
Microstructure noise, 192
Mixture of Distribution Hypothesis, 197
Momentum, 128
Monitoring, 204
Moody's, 61
Moving estimates, 206
Multi-collinearities, 31
Multicollinearity, 128

N

N_2O, 150
NAP, 11, 202

National Balancing Point, 160
Natural gas, 30, 182
Natural gas price, 31, 127
NBP, 160, 163, 166
NCC, 187
Negative covariance, 196
Negative covariance hypothesis, 182, 188, 189, 200
Negative covariation, 187
Net carry cost, 182, 186, 187, 191, 194, 200
New Zealand, 2
New Zealand Unit, 2
Newey-West, 189, 194
News announcements, 20, 27
News arrival, 19
No-arbitrage condition, 152, 186
Nonlinear model, 78
Nonlinear modeling, 74
Nonlinearity, 90, 91, 95
Nonlinearity test, 76
NordPool, 13
Normality test, 110
Number of trades, 200
NYMEX, 13, 170
NZ ETS, 2
NZU, 2

O
Oil, 30, 45
OLS-CUSUM, 114
Option price, 173
Option pricing, 200
Options, 148, 181, 184
OTC, 197
Over-the-counter, 13, 197

P
PACF, 25, 35, 39, 43, 109, 128
PARCH, 65
PCA, 66
Penalty, 7, 12
PFC, 150
Phillips-Perron test, 22, 72
Portfolio frontier, 171, 172
Portfolio management, 13, 170
Portmanteau test, 26, 112, 140, 189, 191, 211
Power sector, 12, 169
PP, 21, 48, 164
Price development, 13
Price dynamics, 14
Price fundamental, 66
Principal component, 67
Principal component analysis, 66, 67

Q
QML, 164, 190
QQ-plot, 24
Quasi-Maximum Likelihood, 142

R
R-squared, 164, 194, 196
RCM, 137, 138
Realized range, 182
Realized volatility, 181, 182, 186, 192, 200
Recessionary shock, 70
Recursive estimates, 205
Regime, 99
Regime classification, 138
Regime Classification Measure, 137
Regime transition probability, 92, 98, 99, 134
Registry, 12, 132
Residual analysis, 109, 122
Residual plots, 24
Residual sums of squares, 28
Residuals, 109
Residuals estimated, 23
RGGI, 2
Risk aversion, 171
Risk factors, 148
Risk management, 183
Risk premia, 151, 153, 156
Risk premium, 181
Risk-management, 181
Risky asset, 61

S
Safe haven, 60
Safety valve, 10
Sampling frequency, 192
Samuelson, 181–183
Samuelson hypothesis, 181, 187, 190, 191, 193, 194, 200
SC, 164, 189
Schwarz criterion, 108, 140
Seasonality, 182, 196, 197, 200
Self-exciting threshold autoregression, 78
Serial correlation, 189, 191
SETAR, 78, 82, 85, 89
Sharpe ratio, 171, 173
Shock, 70
Skewness, 15, 16, 110, 138, 154, 156, 193, 210
Smooth Threshold AutoRegressive Model, 89
Smoothed transition probability, 99, 134
South Korea, 3
Spectral decomposition, 67
Speculator, 183
Spot price, 151, 153, 156
Spot volatility, 200
Spread driver, 131

SSR, 93
Standard error, 24
Standardized innovations, 211
STAR, 87, 89–91, 94, 97, 100
Stationarity, 20, 72, 210
Stochastic, 197
Stock market, 60, 64, 66
Stocks, 60, 61
Storage cost, 63, 66, 153, 154, 156, 189
Structural break, 27, 119, 123, 124, 149, 201,
        203, 204, 210
Structural instability, 73
Student *t* distribution, 177
Switch price, 38, 163, 167
Switching band, 159
Switching point, 36, 159
Systematic risk, 60, 62

T
T-Bill, 60, 62–64, 66, 170, 171
Tangent portfolio, 172
Temperatures index, 41, 42
Terminal condition, 6
TGARCH, 65, 177
Theory of storage, 151, 153, 154
Thermal efficiency, 162
Threshold nonlinearity, 81
Tick-by-tick data, 185
Time-to-expiration, 184
Time-to-maturity, 181–183, 187, 190–192,
        194, 196, 200
Trace statistic test, 210
Trace test, 118
Tradable permits market, 4
Traders, 13
Trading activity, 149
Trading desk, 132
Transactions, 13

Transition function, 88
Transition probability, 134
Transition variable, 94
Tsay test, 75

U
Uncertainty, 184
Unconditional volatility, 181
UNFCCC, 1, 11, 148
Unit root, 48, 81
Unit root test, 164
US Acid Rain Program, 8
US ETS, 2

V
VAR, 67, 107, 141, 211
Variance, 156
VECM, 119, 120, 122
Vector error-correction model, 119
Verified emissions, 12
VIX, 133
Volatility, 45, 133, 149, 181, 183, 184, 201,
        206
Volatility clustering, 56
Volatility signature plot, 192
Volatility-maturity relationship, 184
Volume, 132, 197
Volume of transactions, 15

W
WCI, 2
Weather events, 40
White test, 45, 189
Wold's theorem, 22

Z
Zeebrugge, 31
Zivot-Andrews, 123, 125

Printed by Books on Demand, Germany